Horst-Günter Rubahn

Basics of Nanotechnology

Related Titles

Wolf, E. L.

Quantum Nanoelectronics

An Introduction to Electronic Nanotechnology and Quantum Computing

2009

ISBN: 978-3-527-40749-1

Wolf, E. L.

Nanophysics and Nanotechnology

An Introduction to Modern Concepts in Nanoscience

2006

ISBN: 978-3-527-40651-7

Theodore, L.

Nanotechnology

Basic Calculations for Engineers and Scientists

2005

ISBN: 978-0-471-73951-7

Goodsell, D. S.

Bionanotechnology

Lessons from Nature

2004

ISBN: 978-0-471-41719-4

Poole, C. P., Owens, F. J.

Introduction to Nanotechnology

2003

ISBN: 978-0-471-07935-4

Waser, R. (Ed.)

Nanoelectronics and Information Technology

Advanced Electronic Materials and Novel Devices

2003

ISBN: 978-3-527-40363-9

Horst-Günter Rubahn

Basics of Nanotechnology

Third, Revised and Enlarged Edition

WILEY-VCH Verlag GmbH & Co. KGaA

The Author

Prof. Dr. Horst-Günter Rubahn
University of Southern Denmark
Mads Clausen Institute
Alsion 2
6400 Sonderborg
Danmark

Cover

Atomic force microscopy image of walking stick like, light emitting nanostructures made via self-assembled growth of functionalised organic oligomers on a muscovite mica substrate. Image size: 42 x 42 micrometer squared. Height scale 200 nm.

With permission from M.Schiek, NanoSYD, University of Southern Denmark, Sonderborg, Denmark.

Original Title:

H.-G. Rubahn
Nanophysik und Nanotechnologie
2. Auflage, ISBN 3-519-10331-1
Teubner-Verlag, 2004

Translation:
H.-G. Rubahn

Library of Congress Card No.:
applied for

British Library Cataloguing-in-Publication Data
A catalogue record for this book is available from the British Library.

Bibliographic information published by the Deutsche Nationalbibliothek
Die Deutsche Nationalbibliothek lists this publication in the Deutsche Nationalbibliografie; detailed bibliographic data are available on the Internet at <http://dnb.d-nb.de>.

Composition Da-TeX, Gerd Blumenstein, Leipzig

Printing Betz-Druck GmbH, Darmstadt

Bookbinding Litges & Dopf GmbH, Heppenheim

Printed in the Federal Republic of Germany
Printed on acid-free paper

ISBN: 978-3-527-40800-9

Preface

'Nanotechnology is one of the most important technologies of the twenty-first century'. Such a statement would undoubtedly find support amongst an ever expanding number of scientists, engineers and laymen. The basis of all nanotechnology is 'nanoscience' or somewhat more specifically 'nanophysics', which describes the physics of nanoscaled systems, that is, the transition from atomic physics to continuum and solid state physics.

On such mesoscopic levels between microscopic and macroscopic physics many peculiarities of biology, chemistry and physics vanish. Thus nanotechnology might serve as a very efficient mediator between these disciplines of the natural sciences. The present book tries to take account of this fact – admittedly in a very selective way, biased by the research topics and interests of the author. The book begins with a *physical* approach, describing some of the laws that dictate the possibilities and limits of the new nano-based developments. Technological aspects become important once the processes are described that result in nanostructures, that allow one to characterize them and the ways that make manipulation on a nanoscale possible. *Selforganization* is one of the critical new concepts, a concept which has entered physical thinking on this size range from the chemistry and biology side. The multitude of possible applications in optics, electronics, information theory and biology is illustrated with the help of one-, two- and three-dimensionally nanostructured materials, biological templates and more complex nanomachinery.

This book is an update of a German edition, dating back to the years 2002 and 2004. Since the millenium shift the speed with which nanotechnology has introduced itself into the everyday world of scientists and the rest of the world is simply astonishing. This applies both the use of scientific instruments that base on nanotechnology as well as the appearance of products that are labeled 'nano'. An inofficial survey from 2008 produced over 600 'nano' products, more than 200 of them indeed contain-

Basics of Nanotechnology: 3rd Edition. Horst-Günter Rubahn

ISBN: 978-3-527-40800-9

ing nanoscaled particles or structures. In all research-oriented developed countries the amount of funding for nano-related research is increasing, and also high technology enterprises have discovered the promise of nanotechnology.

In spite of all the technological developments, all the research funding and all the publications in highly ranked journals, basic research breakthroughs that could be termed 'nanotechnology' are still rather scarce – partly because of the extensive optimism with which application writers have been sketching development prognoses within the last ten years. Finally, let us note that the 'Technology Roadmap for Nanoelectronics' (www.cordis.lu) distributed by the European Commission November 2000 is still relevant.

I am deeply grateful to my wife Katharina Rubahn and to my sons Alexander and Markus for the possibility to spend peaceful siesta times on this book. Thanks also to my colleagues Frank Balzer and Jakob Kjelstrup-Hansen for numerous discussions and hints.

Sonderborg, August 2008 *Hoerst-Günther Rubahn*

Contents

Basics of Nanotechnology: 3rd Edition. Horst-Günter Rubahn

ISBN: 978-3-527-40800-9

1 Mesoscopic and Microscopic Physics

According to a generally accepted definition 'nanotechnology' is a technology concerned with objects that have at least in one dimension a size of less than 100 nm. This means that the underlying 'nanophysics' can be placed inbetween mesoscopic physics (the physics of objects between a few microns and one hundred nanometer) and microscopic physics (the physics of interactions between individual atoms and molecules).

Nanotechnology is currently one of the motors driving the introduction of new materials and technologies into all aspects of daily life: from communication and energy generation via health and leisure to traffic and environment. This development can be considered parallel to the present revolution in molecular biology, especially in the molecular biophysics of the post-genomic age. The European Union as well as the large American and Japanese (and to a certain degree also Chinese, Brazilian, Indian etc.) research societies currently invest substantial resources in research geared towards applied nanotechnologies[1]. The main goal for the immediate future is to generate new nano-electronic and nanomechanic elements, to integrate them and to produce the resulting devices in a cheap manner. One of the big expectations is that nanotechnology will eventually fulfill the dreams of scientists from all disciplines; from physicists waiting to see quantum mechanical concepts come to life; via chemists, longing to fabricate large molecules atom for atom; to biologists, seeking to control atom transport into and out of membranes and understand which functions the macromolecules composing the genome perform. *There's plenty of room at the bottom*, as Richard Feynman stated as early as 1959 in his famous lecture [1].

Since the early-1960s the potential of integrated circuits that are used in the computer industry has grown exponentially, fulfilling a prediction

1) Dedicated nanotechnology support through the European Community has been provided since 2000, initially via the 'Nanotechnology information devices'-initiative within framework five (FP5).

Basics of Nanotechnology: 3rd Edition. Horst-Günter Rubahn

ISBN: 978-3-527-40800-9

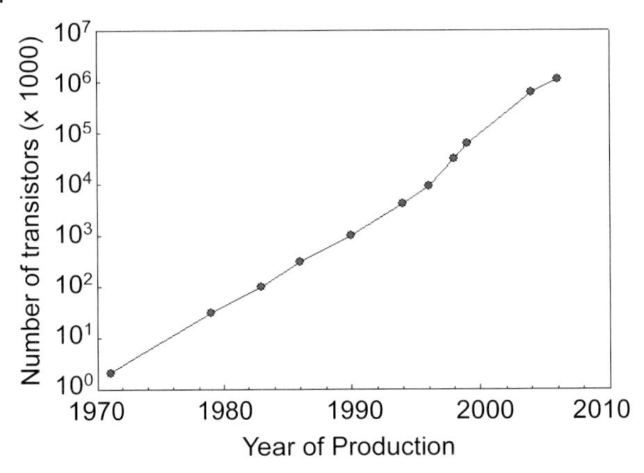

Figure 1.1 Moore's law, confirmed by a plot of number of transistors vs. production year.

by Gordon E. Moore, called 'Moore's law' [2] (Figure 1.1)[2]. Although this 'law' was initially proposed to hold only until the mid-1970s it has since been shown that the physical limit will not be reached until around 2017 [3]. Following a prediction of the SIA ('Semiconductor Industry Association') characteristic structure sizes of 40 nm will be standard in the year 2011 [4]. In fact, Intel currently routinely breaks the 30 nm or even 20 nm size limit. This rapid nano-miniaturization will eventually result in the integration of several tens of million logic elements on a single chip. The bandwidth of information that can be transmitted through an optical fiber doubles even every six months – this made the exponential distribution of the Internet and hence the information technology (IT) revolution possible. New concepts for the preparation of nanostructures, such as neuronal networks, are becoming increasingly important for the traditional, silicon based electronics industry.

Primarily the development has been driven by the microminiaturization of electronic components. This process is mainly limited by technological problems, which can normally be solved by clever engineering. However, approaching atom dimensions, new limits are set by those physical laws which gain increasing importance with decreasing size. This 'physics at small sizes' has the advantage that radically new concepts of information storage and processing can be used, all the way down to the use of quantum mechanical phenomena. In this context new developments such as 'quantum computing' or 'quantum cryptography' gain increasing importance [5].

2) The doubling rate of the number of elements on a chip has decreased from 12 months in the early seventies to roughly 30 months for the first decade of the 21st century.

Although the prefix 'nano' serves in many cases a purely decorative purpose, it has become a synonym for the opening of a new dimension populated with objects that incorporate (or seem to incorporate) new structural, electronic, optical and magnetic properties. The hope is that we will eventually be able to create new materials and processing schemes for our daily 'macroscopic' life: partially via a fundamental understanding (and manipulation) of the microscopic properties and partially via empirical research. The availability of new technological concepts (scanning microscopies, submicron-lithography, laser, super computer) gives hope that these attempts may eventually become successful. Note, however, that the mere reduction in size of participating objects increases tremendously the complexity of the systems to be mastered on a nano-level (Figure 1.2). Present science and technology are still far away from such mastership.

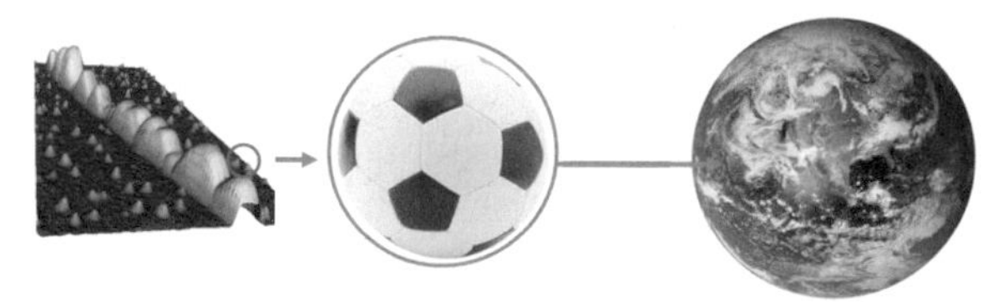

Figure 1.2 A football filled with nanoscaled objects resembles the complexity of the whole earth.

Taking into account all the above mentioned developments, it is nearly unavoidable to fall into a state of technological hybris. Let us, therefore, remember that many of the 'newly developed' nano effects have historical roots. Glasses, for example, which show extreme brilliance by the inclusion of colloidal quantum dots, have been know since ancient Greek times: from the famous Lycurgus cup to church windows. In these cases the colloidal solutions were generated in 400 B.C. via 'alchimia' and the quantum dots fabricated by the mechanical formation of gold and silver dust particles.

The most common method for the fabrication of structures in the submicrometer size regime is the 'top-down' technology ('from large to small'), where using lithographic techniques nanoscaled elements are cut from larger entities (Figure 1.3). However, as one approaches sizes below 100 nm the resolution and replication speed limit the efficiency, consequently in this size regime the 'bottom-up' technology ('from small to large') becomes important. Here nanosized objects are formed from their atomic or molecular compounds using appropriate building recipes. Two different approaches are available. In the first one non-biological molecular mechanisms are used on a surface or in a liq-

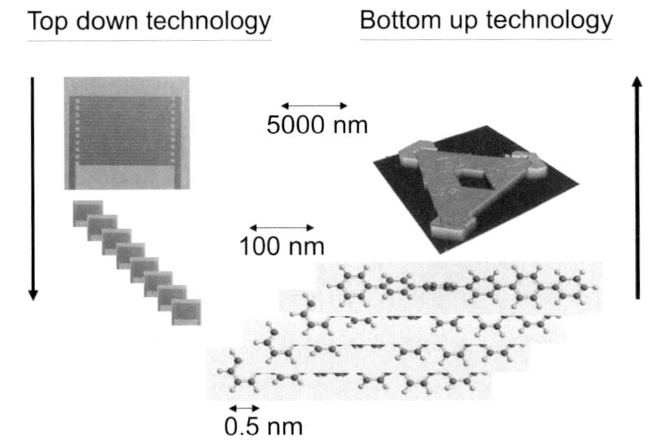

Figure 1.3 Top-down vs. bottom-up technology.

uid to run controlled chemical (surface) reactions that produce complex structures that are specified down to the atom scale. The results of such mechanistic engineering science on the atom level is a non-biological[3] 'molecular manufacturing' or 'molecular nanotechnology' [7, 8], which contains as its most important elements mechanical driven chemical synthesis (e.g. direct positioning of reactive molecules on a surface), energetically driven surface structure formation and/or molecular transformation of chemical into kinetic energy (e.g. via synthetic molecular actuators and motors).

It is worthwhile to compare top-down fabrication, such as conventional macroscopic or microstructure fabrication, with molecular manufacturing [7]. While the feature scale decreases from 1 mm (macroscopic fabrication) over 1 μm (microscopic fabrication) to 0.3 nm (molecular manufacturing), the product scale decreases from 1 m over 10 mm to about 100 nm, but at the same time the defect rate per component also decreases drastically from 10^{-4} over 10^{-7} to 10^{-15}. This is accompanied by an increase in cycle times from 1 s to 100 s. However, cycle times for molecular manufacturing are only of the order of 1 μs.

It is also possible to direct a self-consistent system development in a mimicry of biological development on an atomic or molecular level. This principle of 'self organization', driven by conversion of free energy [9], is not limited to biological systems[4] but can be applied to a large va-

3) Molecular nanotechnology has a rather wide range of implications, for example for a new kind of medical diagnosis and therapy[6].

4) Biology is limited to the aqueous phase and is concerned with 'soft' materials with characteristic energies of phase change of $k_B T$. Even small temperature changes result in huge struc-

riety of atomic and molecular architectures. Another advantage of the 'bottom-up' approach as compared to the 'top-down' approach is that it is a massive parallel approach. Every product mol unit contains some 10^{23} individual yet identically built nano systems.

A basic question to every approach on nanotechnology is: in how far do nanoscaled objects behave in the same way as objects in the macroscopic world, just on a smaller size scale. In other words: how reliably can one apply physical or chemical laws of the macroscopic world on objects that consist of a countable number of atoms and have distances from each other of the order of countable multitudes of atom diameters? It is found, for example, that in the nano world the equilibrium of forces is different from that in the macroscopic world: owing of the small mass of the objects, gravitational forces are less important while electrostatic attraction and van der Waals forces are of huge importance. The electrostatic force is about 11 orders of magnitude larger compared to the magnetic force between two parallel, 1 nm long conductor segments through which a current of 10 nA flows. This latter force is only $2 \cdot 10^{-23}$ N and thus much smaller than the covalent bond strength which is of the order of 10^{-9} N[5]. The small dimensions of the nano-objects and their low masses make fast frequency motion possible ($100\,\mathrm{ms}^{-1}$ velocity mean that 1 nm is traversed in 10 ps). Pressure values are large for small forces due to the small areas involved – this is also true for light pressure. Photons can redirect easily non-bound nano objects.

If one aims to move nanoscaled objects to a defined location with a defined speed, then in additon to the quantum mechanical uncertainty limit the statistical movement of the nanoobjects has to be taken into account. This statistical movement is induced by the temperature of the environment ('Brownian movement'). The thermal velocity is proportional to the square root of the ratio between thermal energy and mass. For one cubic nanometer diamond (density $3.5 \cdot 10^3\,\mathrm{kgm}^{-3}$) one obtains at room temperature an average thermal velocity of $60\,\mathrm{ms}^{-1}$, that is, $60 \times 10^9\,\mathrm{nms}^{-1}$.

In general, the ratio between thermal energy kT and quantum mechanical energy quantum $\hbar\omega$ determines whether statistical uncertainties due to thermal fluctuations or quantum mechanical uncertainties due to the quantum mechanical uncertainty principle are the dominant factors that limit the ability to localize position and velocity of a nanoscaled object with high precision.

tural changes. Self organization, however, can be found also in physical or chemical much stronger bound systems,e.g., within supramolecular chemistry.

5) Note, however, that these values result from classical continuum scaling laws, which might not represent accurately the nano world.

During the interplay of two objects on the nanoscale one has to take into account that classical lubricants no longer work as expected since fluids loose their viscosity close to surfaces and in many cases no longer behave as fluids; consequently, friction might lead to strong wear and thus to a short lifetime of nanoscaled mechanical machinery. However, if one looks closer at the problem in terms of attractive and repulsive molecule-surface forces instead of the macroscopic description via friction one finds that the resulting equilibrium of forces can significantly decrease wear and thus increase characteristic lifetimes. The lesson being that true nanophysical effects make simple estimates obsolete – both in the positive and in the negative directions.

Finally, it is noted that the 'nano revolution' is not limited to the structural dimension – the interaction *dynamics* on the atom level are just as important. Movement and movability of atomic, molecular and nanoscaled objects have to be investigated and understood before one can confidently manipulate them on a nanoscale. An important tool for such are ultrashort pulse lasers ('femtosecond' lasers) which allow one to image and manipulate elemental dynamics with atom resolution.

The uncertainty relation tells us that a light pulse with a duration of some ten femtoseconds (10^{-15} s) has a huge energy uncertainty of some electron volts (eV) (Figure 1.4a). Since the energetical distance between vibrational Eigenstates of a molecule is far below one eV, the excitation of the molecule with the femtosecond pulse results in a coherent superposition of excited Eigenstates. Further, this uncertainty in the energetic excitation means that the Eigenstates of the molecule are transformed from their original spatial distribution along the space coordinate to become localized within the femtosecond excited state into a wave packet with small spatial extension (Figure 1.4b). The movement of this wave packet along a potential curve can thus be registered with sub-Ångstrom resolution.

In the recent past this technology has also been applied to problems of biological relevance ('femtobiology'). One investigates, for example, conformational changes in proteins on a realtime scale or monitors temporally resolved electron transport via DNA. From the latter measurements it has been deduced, among others, that DNA possesses only a small conductance, similar to a semiconductor. This implies that the direct use of DNA as a 'molecular wire' in future quantum computers is not possible.

In summary, modern technology allows one to explore the nano-world with extremely high spatial and temporal resolution. However, there is still a long way to go from a detailed exploration to a complete understanding or directed manipulation.

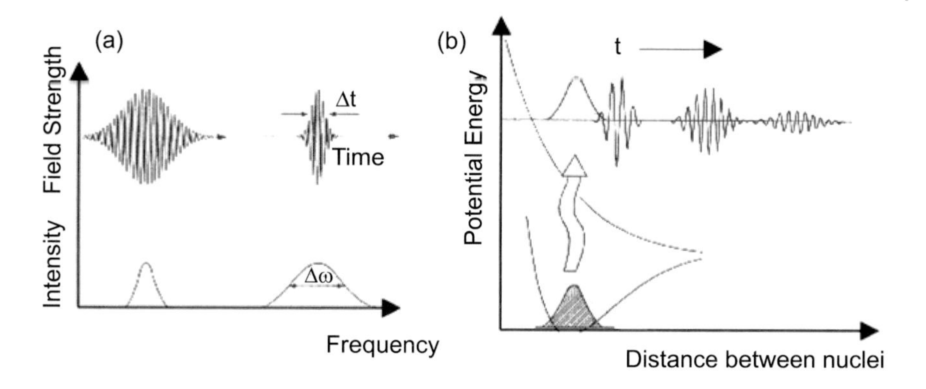

Figure 1.4 (a) The relationship between temporal and energetical uncertainty. (b) Generation of a localized wave packet via excitation with a femtosecond pulse. Printed with permission from [10]. Copyright 1993 Cambridge University Press.

Problems

Problem 1.1 Compare gravitational, electrostatic and van der Waals forces between two 100 nm diameter polystyrole spheres which are separated by a) one micrometer, b) one millimeter and c) one centimeter.

Problem 1.2 The momentum of a photon is $\hbar k$, where k is the wave vector. Assume an object moving at a speed of $10\,\mathrm{ms}^{-1}$. Objects of which mass can be stopped by elastic reflection of photon?

Problem 1.3 Ultrashort pulsed lasers allow one to investigate dynamics in the nano world with unprecedented temporal resolution. What is the minimum temporal length of a laser pulse of wavelength 800 nm?

Problem 1.4 A typical femtosecond laser pulse has a duration of 80 fs and a central wavelength of 800 nm. Calculate the corresponding spectral line width, that is, the uncertainty in energy of the photons. The temporal resolution allows one to study very precisely dynamics in nanoscaled systems, but is the corresponding energy definition sufficient to resolve optical transitions between a) electronic states, b) vibrational states, c) rotational states of molecules? Use as an example the lithium dimer and the electronically excited B-state. Term energy is $21926\,\mathrm{cm}^{-1}$, vibrational constant $351\,\mathrm{cm}^{-1}$ and rotational constant $0.67\,\mathrm{cm}^{-1}$.

2
From Isolated Atoms to the Bulk

Nanotechnology aims to miniaturize devices down to the atom scale. In both top-down and bottom-up strategies it will become necessary to solve problems of a mesoscopic nature, especially concerning the transition from the solid bulk to the atom and vice versa: nanometer sized aggregates, built up from individual atoms, have to have a lattice structure and an electronic band structure in order to be able to treat the electrons theoretically as a collective unit. This transition from electrons that are affiliated with specific atoms to a collective 'jellium' or gas of electrons around fixed ionic cores can occur at very small sized aggregates[1]. However, in other cases it might also require a very large number of atoms. An important aspect of modern 'cluster research' is to understand quantitatively the details of this process. In general, one observes that the material properties of the clusters as well as their environment play an important role.

A thorough understanding of the transition from atom to bulk properties enables one to take advantage of the extraordinary change of properties which particles experience during this transition. This concerns not only optical properties as described in Section 2.3 but also magnetic and thermodynamic properties such as the melting temperature, which decreases with decreasing size, or the pressure values that are necessary for phase transitions, which increase with decreasing size. Besides quantization effects, the increasing number of surface atoms as compared to the bulk atoms also specifies the behaviour. Since surface atoms determine the free energy of the particles, a change in their relative number with respect to bulk atoms will lead to a change of the thermodynamic properties of the particle.

1) Gold clusters with the 'magic' number of 55 atoms (a completely packed shell with 12 and one with 42 atoms around a central atom) and a diameter of 1.44 nm possess metallic properties, pointing to the existence of a collectively excitable electron gas [11]. Even in very small semiconductor nanocrystallites one observes well defined single crystalline facets.

Basics of Nanotechnology: 3rd Edition. Horst-Günter Rubahn

ISBN: 978-3-527-40800-9

2.1
Matter and Waves

Particle systems or even devices with length scales of the order of the electron de Broglie wavelength certainly have to be treated in the quantum mechanical limit, taking into account the dualism of particle and wave. Essentially, a particle of mass m travelling with velocity v and momentum mv can be associated with a wave with frequency ν and energy $h \cdot \nu$ that has a corresponding wavelength

$$\lambda_{dB} = \frac{h}{mv} \quad , \tag{2.1}$$

the de Broglie wavelength. The position of such a particle is thus given by the probability $\psi^* \psi$ to detect a corresponding wave package $\psi(r,t)$ at time t and coordinate $\vec{r}$. The wave package is a superposition of plane waves with distribution functions in, for example, location (in one dimension: Δx) and momentum Δp related via the de Broglie wavelength as

$$\Delta x \Delta p \simeq \hbar. \tag{2.2}$$

This is called one of the 'uncertainty relations' in quantum mechanics. Since statistical probabilities are what can be observed in reality, a precise definition of one of these observables (e.g. location) results unavoidably in an unsharp definition of the corresponding observable (momentum).

In analogy to classical mechanics an equation of motion can also be written for quantum mechanics, which is called the Schrödinger equation:

$$i\hbar\dot{\psi} = \hat{H}\psi \quad , \tag{2.3}$$

which relates the differential change in time t to a change in the energy observable that is described by the Hamilton operator $\hat{H}$. The Hamilton operator includes the kinetic and potential energy of the system and thus depends on the forces that are acting on the particle (e.g. gravitation, electric or magnetic forces etc.). Solutions of this differential equation result in the Eigenfunctions and Eigenvalues of the system, for example the characteristic wavefunctions that span an optical potential or the transition energies between discrete states inside of this potential.

2.2
Morphology

Understanding the structure and mechanical properties of nanostructured materials is important for the manipulation and modification at the atomic level. Increasing the number of atoms from a single one via a cluster to the bulk, one finds 'magic numbers' which correspond to crystalline structures of high stability. These are reflected, for example, in peaks in the appearance rate of clusters of different sizes. In the case of rare gas clusters with an isotropic atom interaction potential, the stabile configurations correspond to a dense packing of atoms, for example, icosaeders around a central atom with increasing particle number ('magic numbers' of 13, 55, 147 ...). With increasing size these five-fold symmetries show more and more defects: in the case of argon one observes at around 1000 atoms the crystalline structure of the bulk solid.

For more complex atoms (e.g. metals) with anisotropic interaction potentials or open shells the morphologic rules become more complicated. For large clusters magic numbers exist, but for smaller clusters electronic effects dominate [12]. In this range of smaller clusters the closed electronic shells (N=8, 18, 20 ...) result in most stable clusters.

In fact it would be surprising to observe 'simple' morphologic rules since crystal structures are affected by small changes in the long range interaction potential. For aggregates with dimensions of the order of several atom distances, a strong influence of these 'surface' effects is expected and the stability of such an aggregate cannot be easily predicted.

Besides just calculating or measuring the resulting structures for a given number of atoms, one might also be interested in deducing the structure of a nanoscaled aggregate of given dimensions and calculating whether such a structure is stable or not. In this context the most important structures are connecting elements between specific aggregates, for example, simple wires. An important question is how thin could such an atom wire be and how stable would it be?

The mechanical properties of a wire of individual gold atoms have recently been investigated by a combination of scanning tunneling microscopy (STM) and an atomic force (AFM) sensor [13]. The gold wire was connected between the force microscope tip and the golden tip of an STM. The force necessary to stretch the wire was then determined with the help of a second STM which detected the extension of the AFM tip. The somewhat surprising result was that the bond strength between atoms in the nanowire is twice that of the bond strength in the bulk. Apparently the effective stiffness of a nanostructure depends on the exact configuration of the atoms at the basis.

2.3
Electronic Structure and Optical Properties

Quantum mechanical size effects are expected if the size of an object approaches the characteristic length scale which determines the coherence of its wavefunction. At or below this size the electronic, optical and magnetic properties become size- and shape-dependent, for all the optical behaviour is very sensitive to quantum effects ('quantum confinement'). The significance of the effects depends on the temperature; namely, the distance between neighboring energy levels or bands has to be large compared to the thermal energy $k_B T$ in order to avoid a smear-out of the effect by thermal fluctuations.

The absorption of a photon results in an electron-hole pair in the material, more exactly a quantized state called an exciton. In semiconductors because of the small effective masses of electrons, m_e, and holes, m_h, ('effective' meaning the masses in the presence of the lattice) and the large dielectric constant ϵ, the exciton is large on the atom scale, with Bohr radii a_0 between 5 nm and 50 nm. The Bohr radius of an exciton is given by

$$a_0 = \frac{\hbar^2 \epsilon}{e^2} \left[\frac{1}{m_e} + \frac{1}{m_h} \right] \quad . \tag{2.4}$$

If the semiconducting aggregate reaches in one (quantum gas), two (quantum wire) or three dimensions (quantum dot) a size that is comparable with a_0 or the de Broglie wavelength λ_{dB} (Equation 2.1) massive changes in the optical properties are observed. In metals such effects are observed only for very small aggregates of a few nanometers on account of the missing band gap and the corresponding small distances between relevant energy levels. In Figure 2.1 the change of electronic density of states for a semiconductor band is plotted as a function of reduced

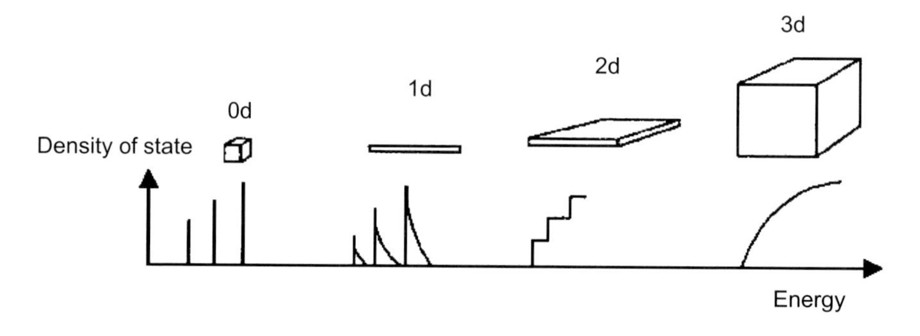

Figure 2.1 Electronic densities of state for semiconductor nanostructures of various dimensions.

dimensions. The quantum dot ('zero dimensions') has discrete energy levels, the quantum wire ('one dimension') quasi-discrete and the quantum well ('two dimensions') a quantized stairs pattern. The bulk has a continuous spectrum.

2.3.1 Semiconductors: Quantum Dots and Wires

Semiconductor quantum dots can be fabricated via epitaxial growth or with the help of synthetic chemistry from colloidal solutions (cf. Section 5.2). The typical surface density of self organized quantum dots is $10^{10}\,\mathrm{cm}^{-2}$ to $10^{12}\,\mathrm{cm}^{-2}$. Hence, we can call a matrix of quantum dots a two-dimensional array of 'artificial atoms' which possess a narrow distribution of excitation energies [14].

How one can change optical properties by simply changing the sizes of nanoaggregates is most easily demonstrated in colloidal solutions of quantum dots. Such solutions with semiconductor quantum dots (e.g. CdS) of average radius 4.1 nm emit red light, which varies as a function of size from yellow (radius 1.7 nm) to green (radius 1.2 nm) (see [15] for a collection of microphotographic images).

Since the ground state of a quantum dot is not degenerate, a single exciton per dot is sufficient for population inversion of the energy levels. Epitaxially grown or solution based matrices of quantum dots (zero-dimensional excitons) therefore represent as successors of quantum well lasers (two-dimensional excitons) ideal media for laser activity with minimum threshold current densities. Recently, size-controlled stimulated emission from quantum dots has been demonstrated [16]. Laser light from the blue to the red spectral range has been generated by the use of 2 nm to 10 nm sized quantum dots [17].

With the help of single molecule spectroscopy it becomes possible to investigate the optical dynamics of isolated quantum dots. The diffraction limited resolution of confocal microscopy is about $\lambda/2 \approx 250\,\mathrm{nm}$. If one reduces the surface density of quantum dots to one dot per square micrometer one works indeed with isolated quantum dots. Note also that quantum dots are significantly more stable than, for example, dye molecules, which bleach after a few 10^6 excitation-emission cycles. Quantum dots are thus perfect model systems for the investigation of photodynamics in confined systems.

An interesting effect is a 'blinking' of the quantum dots, which happens on timescales between 10^{-4} and 10^3 s. A possible reason for this blinking is electron/hole pair ejections and recombination dynamics. In the 'on'-state electrons from the core of the quantum dot are excited, in the 'off'-state the electrons are far away from the holes and in the next 'on'-state they recombine again. Electron transport occurs via tunneling and from the measured lifetime a minimum distance between hole and electron of 2 – 4 nm can be estimated. Kinetics obeys a power law ($P(t) \propto 1/\tau^m$), since a localized excitation including a single exponential lifetime dependence is weighted with an exponential distribution of rate constants (distribution of potential well depths in the excited state of the quantum dot). Thus, this kind of dynamics cannot be described by an average lifetime but by a distribution of lifetimes. In other words: the 'lifetime' of the blinking is dictated by the experimental details since one always finds events which occur on a time scale that is either too long or too short to be observed. The process is intrinsically dynamic and fluctuations in the nano-environment of the quantum dot play an important role [18,19].

Upon increasing the size of the quantum dots to quantum wires one will find above a certain aspect ratio (> 2) a preferred axis (c-axis), which defines a preferred direction of the transition dipole moment. From this point on excitation and emission characteristics are strongly linearly polarized [20]. Thus, structure has a strong influence on the static and dynamic optical properties of these nanoscaled systems.

2.3.2 Metals

The spectra of isolated alkali atoms, -dimers and -trimers are determined by single electron excitations. They consist – measured under appropriate conditions[2] – of well defined spectral lines and can be in theory, quantitatively reproduced. Figure 2.2 shows as examples a Doppler-free two-photon fluorescence spectrum of sodium atoms and single photon fluorescence of sodium dimers. The ordinate in Figure 2.2a is a relative frequency difference between the two transitions. For Figure 2.2b absolute wavelengths of the transitions between individual rovibronic states of the molecule are given.

The character of the optical spectra changes with increasing number of atoms in the polymer ('cluster'). In the case of sodium, for clusters of

2) Appropriate conditions are those under which the particles possess a well-defined single velocity. If they, however, obey a velocity distribution the observed linewidth will be dominated by the Doppler effect. In a thermal distribution this Doppler broadening is 100 to 1000 times larger compared to the natural linewidth, which is given by the lifetime of the upper excited electronic state. The Doppler broadening due to the fixed velocity of the particles can be reduced via optical excitation perpendicular to the velocity vector of the particles.

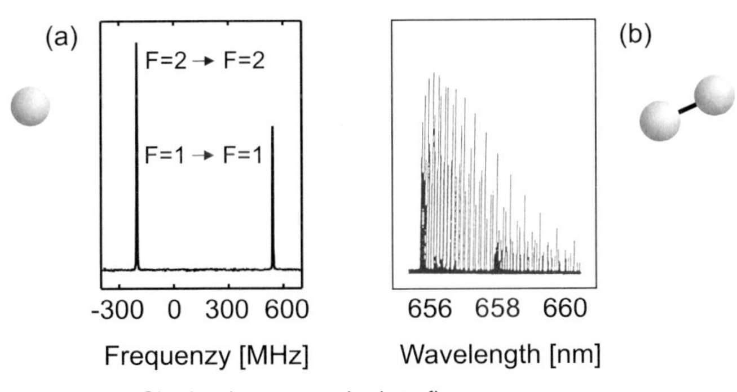

Figure 2.2 Single electron excitation: fluorescence spectra of sodium atoms (a) and sodium dimers (b). The corresponding term schemes are shown in Figure 2.4.

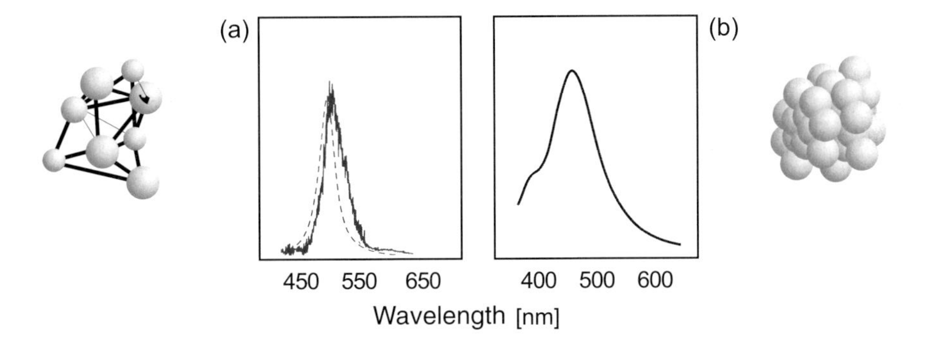

Figure 2.3 Collective electronic excitation: absorption spectrum of Na_8 clusters (a) and extinction spectrum of large, surface deposited clusters (b). Image (a) reprinted with permission from [21]. Copyright 1990 American Institute of Physics.

eight atoms the oscillator strength is concentrated on a single absorption maximum (Figure 2.3a) [21,22]. Instead of single electron excitation a collective excitation of the valence electrons is observed, a 'surface plasmon excitation'. The particle size at which this transition from localized to collective excitation occurs depends intrinsically on the kind of atoms in the charged state. For example, Na_{10}^+ ions show well separated photoabsorption lines, whereas Na_{15}^+ ions have photoabsorption spectra, which can be described via elemental excitations of the free electron gas [23].

This delocalization of the conduction band electrons is also found in much larger clusters, even if they are adsorbed as nanoparticles on surfaces (Figure 2.3b). In such cases it is no longer possible to address individual atoms in these aggregates with optical measures. On the other hand, however, it becomes easier to reproduce quantitatively the optical

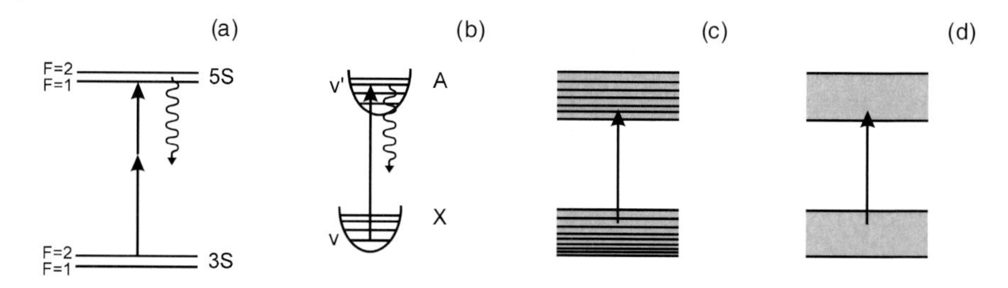

Figure 2.4 Term schemes of optical excitations, which lead to the spectra shown in Figures 2.2 and 2.3. The ordinate is the potential energy.

behaviour of the aggregates since the electronic excitations can be described to very good approximation using the 'jellium' model; namely, the electrons fill as a structureless 'jelly' the space between the rigid ion cores. A quantitative description of the nano-optics of these clusters in turn makes it possible to obtain morphological information using optical measurement tools (see also Section 4.3).

A thorough understanding of both the optical response as well as the structural stability of the clusters becomes difficult in the transition regime between isolated atoms and very large clusters. One observes, for example, even for surprisingly small Na_n^+ clusters ($n < 93$) that the behaviour of the particle resonance as a function of cluster size is very similar to the dispersion relationship of an infinitely extended surface composed from the same material [24]. The exact size dependence and the lifetime of the plasmon resonance which determine the optical response are topics of intense experimental [25–27] and theoretical investigations [28–31] (see also Section 6.1.4).

If one changes the geometrical structure of the metallic quantum dots by expanding them anisotropically to quantum wires, the electronic and optical behavior changes by a similar amount as in the case of semiconductor quantum wires. More details about the electric conductance of atom quantum wires can be found in Section 6.2.2.

Problems

Problem 2.1 Location and momentum of a quantum mechanical particle cannot be determined simultaneously with the same precision. If you determine the energy, which other quantity cannot be simultaneously precisely determined?

Problem 2.2 The energetic positions of the line maxima in Figure 2.3a and Figure 2.3b are nearly the same, although the metallic cluster size changes from around 1 nm to 40 nm. In contrast: Why does the emitted light wavelength depend so strongly on the size in the case of semiconductor quantum dots?

Problem 2.3 Which photon energies are used for the transitions shown in Figure 2.2b and Figure 2.3b? Why are the particles not destroyed if one pumps that much energy into them? For example, atoms can be desorbed from the surface of the clusters shown in Figure 2.3b via thermal excitation. Which energy is associated with a thermal heating of 800 K?

3
Generation and Manipulation of Nanostructures

Nanostructures can be fabricated either via micro-miniaturization with advanced technologies ('top-down methods') or through inducing the conditions for self organized growth of the structures ('bottom-up method').

3.1
Top-down Methods

Fabrication of nanostructures of arbitrary shape has been possible for over 40 years [32]. In principle the structures can be generated in or on a multitude of materials via projection methods [33] or via direct-write approaches. Both approaches have their advantages and disadvantages. The *projection methods* use a mask, which is imaged via photons, electrons, ions or atoms on a light- or particle-sensitive substrate. This results in the easy production of large numbers of similar structures. The minimum structure size is determined by diffraction limits during exposure with photons or particles. In addition, the lifetime of the mask is rather limited due to radiation damage – this in turn results in a deterioration of the imaging quality. The *direct write methods* use in most cases electrons or ions. Due to the short de Broglie wavelengths of these particles one can fabricate relatively simple structures with a minimum size determined by the scattering processes in the material being treated (proximity-effect). Disadvantages are the large focal depth and the time required to fabricate the structures via scanning along predefined patterns. In most cases one produces via direct write methods the masks, which in a subsequent step are used for projection multiplication.

3.1.1
Nanostructures via Photons and Lithography

In this paragraph we discuss lithographical techniques which use light-, electron- or ion-mediated imprinting of structures into solid state sur-

Basics of Nanotechnology: 3rd Edition. Horst-Günter Rubahn

ISBN: 978-3-527-40800-9

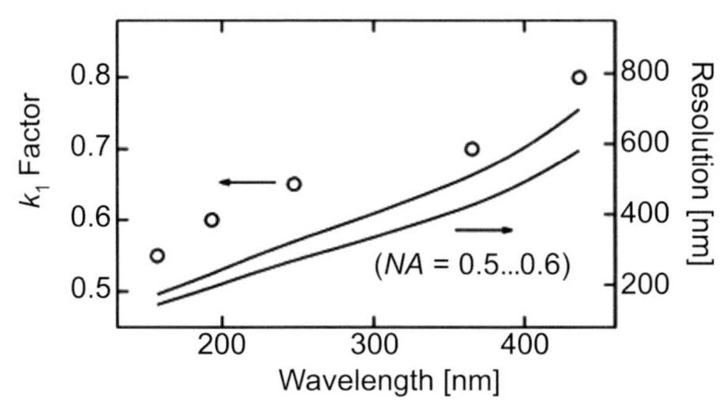

Figure 3.1 Coherence factor k_1 (open circles [35]) and range of possible lithographical resolutions for numerical apertures between 0.5 and 0.6.

faces. It is noted that it is possible to generate nanoscaled structures in the gas phase via interfering light waves. The manipulation of free particles with light forces is described in Section 5.3.

Imaging Photolithography

The minimum structure size d_{min} which can be obtained in photolithographic methods scales with the wavelength λ and is given by the Rayleigh criterion (Equation 4.5). For a spatial resolution below 100 nm one has to use smaller imaging wavelengths (e.g. $\lambda = 157\,\text{nm}$ from an F_2 excimer laser) [34]. Handling this ultrashort wavelength radiation is, however, anything but trivial. Due to strong absorption in ambient air the radiation path has to be purged with dry nitrogen. An additional problem are color centers[1] which are generated inside the initially only very weakly absorbing dielectric lens materials ('laser damage'). At these color centers enhanced absorption occurs, which results in material removal. Since there are no materials without defects this process limits the lifetime of all optical components. Even for very good elements the average lifetime is far less than ten years.

Besides wavelength, the numerical aperture NA and the coherence factor k_1 in Equation 4.5 limit the possible resolution. In Figure 3.1 the calculated lithographical resolution is plotted for values of $NA = 0.5$ to $NA = 0.6$. The coherence factor k_1 can be approximated by a linear regression of the form $k_1 = 0.44 + 8 \times 10^{-4} \cdot \lambda$ [nm].

An increase in the numerical aperture (more partial waves contribute to the image) results in a higher resolution. However, the focus depth f_{min} decreases with increasing NA and thus also the depth of the struc-

1) A color center is, for example, a missing ion in an ionic crystal such as KCl, which traps an optically excitable electron.

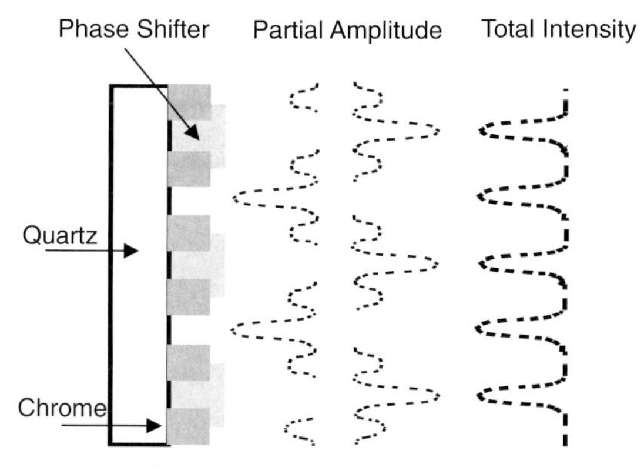

Figure 3.2 Principle of a phase shift mask. The amplitudes of the electromagnetic field of neighboring transparent areas are phase shifted by 180°, thus reducing the diffraction intensity of the dark areas and in turn increasing the effective contrast.

tured area:

$$f_{\min} = k_2 \frac{\lambda}{NA^2}, \tag{3.1}$$

with k_2 a geometrical factor[2] with a similar value as k_1. Thus the parameter that is most akin to optimization is k_1. The size of this factor can be reduced, for example, via phase shift masks [37, 38]. Let us assume that we want to image a grating via coherent light. Due to diffraction from the bright to the dark regions the quality of the image decreases. This can be avoided by introducing a phase shift layer with refractive index n on every second transparent region. The layer has a thickness

$$d = \frac{\lambda}{2(n-1)}, \tag{3.2}$$

and shifts the phase of the light by 180°, that is, leads to destructive interference in the dark areas.

In real systems, however, the resolution is also limited by wavefront aberrations between interfering partial waves, which result from the fact that diffracted beams of zeroth and higher order possess different optical paths. This problem can be solved, and thus resolution and focus depth can be improved, if one blocks the central part of the initial beam; at the same time rays of zeroth and first order should interfere in the image plane under the same angle of incidence ('annular illumination' [39]). In combination with 'soft' attenuated phase shift masks 'off-axis-illumination' results in even higher resolution, characterized by k_1

2) In the Rayleigh limit one finds k_2=$k_1/2$ [36].

values up to 0.25 [40]. A more extended introduction into these kind of problems in microlithography and microfabrication can be found in [41].

Electron Beam Lithography

In electron beam lithography structures are written in an electron resist (usually a polymer such as PMMA, polymethyl methacrylate) via scanning of a focused high energy electron beam (typical 20 keV). This resist structure acts as a mask for photolithography. 'Writing' in terms of electron irradiation means that the structure of the resist is changed at the irradiated place (e.g. bonds are broken and the molecular density is changed correspondingly) such that in a subsequent process the irradiated area can be dissolved in a chemical reagent. Depending on the structure to be etched, positive and negative electron resists are used.

The minimum resolution is limited by the proximity effect, that is, the forward and backward scattering of electrons inside the resist. Scattering increases the width of the illuminating electron cone (by up to a few micrometers), which means that the electrons also irradiate areas of the resist a few micrometers outside the axis of the incoming electron beam. Therefore, small or isolated structures obtain a smaller amount of electrons compared to larger structures and so require for their development a correspondingly larger dose. In Figure 3.3 this is demonstrated by a narrow line in the immediate neighborhood of a wide line. The narrow line obviously is less well developed.

In order to avoid this effect, a modulation of the electron dose is necessary. This is easily regulated for structures with uniform density and line width (e.g. isolated transistor structures). In the case of inhomogeneous structures the dose is varied during scanning. This requires an at least approximate calculation of the necessary dose before the scanning process. Using the GHOST method [43,44] such a calculation is not necessary. Here, an inverse structure image is written with the help of a defocused beam – this image mimics the backscattering distribution. The (positive) electron dose is chosen such that together with the dose of the defocused beam the optimum dose for the development of all structures is obtained. This method requires an extensive calculation of the inverse scattering distribution and the fabrication of a complex mask with several amplitude- and phase-levels.

Besides minimizing the backward scattering, forward scattering also has to be eliminated in order to avoid electron beam defocusing. This can be achieved by a multilayer resist which consists of a thin cover layer that is sensitive to the electron irradiation; the structure developed in this layer is then transferred into a thicker, supporting layer by dry etching.

Figure 3.3 Proximity effect: Scanning electron microscopy (SEM) image of a pattern in silicon, scanned with 20 kV electrons. The electron dose was kept constant over the whole surface area. The narrow structures received less dose compared to the wide ones and are thus less well developed. Reprinted with permission from [42]. Copyright 1981 American Institute of Physics.

In Table 3.1 the minimum achievable structure sizes d_{min} and minimum focus depths f_{min} are listed for some lithographical techniques [45] such as ion beam, X-ray- and UV-laser based techniques (193 nm and 157 nm). 'SCALPEL' [46] stands for *Scattering with Angular Limitation Projection Electron-beam Lithography*[3] and 'EUV' for *Extreme ultraviolet*.

Table 3.1: Overview of minimum structure sizes and focus depths of some lithographic techniques. 'em' means that electromagnetic optics is used, 'refl.' that reflection and 'trans.' that transmission optics are implemented.

Technique	λ	d_{min} [nm]	f_{min} [μm]	Introduced (year)	Optic Type
ion beam	50 fm	2	500	80s	em
SCALPEL	4 pm	0.16	400	80s	em
X-ray	1 nm	30	–	70s	refl.
EUV	11–14 nm	45	1.1	80s	refl.
193 nm	193 nm	100	0.4	80s	trans.
157 nm	157 nm	80	0.28	80s	trans.

3) In SCALPEL a mask for electron beam imaging is used which separates contrast formation and energy absorption into two different layers: a structured, weakly absorbing membrane layer covers a strong scattering, non-structured layer.

Phase shift technology allows one to obtain effective structure widths of, for example, a transistor basis between 20 nm and 50 nm even with conventional wavelengths of 193 nm or 248 nm [45]. The wavelength of the applied light and the resulting diffraction phenomena are thus not necessarily limiting in respect of a further increase of transistor density in integrated chips.

Unconventional Lithographic Techniques

In recent years a multitude of methods have been developed for the generation of periodic structures from metallic and semiconducting [47–50] materials on a nanometric scale, methods which go beyond conventional lithography. Examples include:

Nanosphere lithography [51–53]
Here the selforganization of colloidal spheres is used for lithography in the submicrometer range. A solution of polystyrole spheres with diameters ranging from a few microns down to a few hundred nanometers is sprayed onto a substrate (e.g. a single crystalline silicon or a simple microscope slide) and thereafter dried out. Due to capillary forces the spheres attract each other and form an ordered monolayer. In a subsequent step metal (silver, gold or chromium) is evaporated on top and the polystyrole spheres are dissolved in CH_2Cl_2. As a result one obtains ordered distributions of triangular (using a monolayer polystyrole), spherical (using a double layer polystyrole) or ring-shaped [54] nano particles.

Laser focussing [55,56] *and interference lithography* [57]
Atoms from a thermal source pass during surface deposition through a standing laser wave with wavelength λ, which is quasiresonant with an electronic transition in the atom. Due to induced dipole moments a force is exerted onto the atoms in the direction of the field strength maxima of the standing wave (see also Section 5.3). Depending on the pattern of the standing wave parallel, metallic lines (sodium, chromium, aluminum) can be generated with widths below 30 nm and distances of $\lambda/2$, but also periodic patterns of isolated clusters.

As an alternative one might generate an interference pattern by overlaying two UV laser beams on a photo lacquer – this leads to a structured illumination of the lacquer. Following conventional development one can grow periodic arrays of clusters from more or less all materials at the illuminated places. Using this method large area periodic magnetic nanostructures with periodicities down to 130 nm have been fabricated [57]. The size of the individual cobald/platinum double layer quantum dots is of the order of 70 nm.

Atom lithography [58]
Similar to laser focussed atom deposition in this process one uses the forces that near resonant laser light exerts on alkali atoms or metastable rare gas atoms to write periodic structures with widths far below 100 nm. Instead of depositing atoms directly on the surface, ultrathin organic films are modified chemically through interaction with the metastable rare gas atoms or the alkali atoms (especially cesium). In a subsequent wet etching step these irradiated areas can be removed [59,60].

Metastable rare gas atoms also depassivate hydrogen-passivated silicon [61] such that one can write structures on those surfaces. Characteristic structure sizes are determined by the laser beams, which exert forces on the metastable atoms. An important advantage compared to the direct deposition of alkali atoms is that the structured surfaces are chemically inert and so further layers can be deposited on top.

Laser Ablation

A well established method for the generation of surface structures in the micrometer- or submicrometer-range is UV laser ablation [62]. With this method even diamond with its very high enthalpy of evaporation and thermal conductivity can be selectively manufactured. If one applies nanosecond pulses, a carbon layer is formed around the ablation hole owing to the large conductivity. Ultrashort (femtosecond) pulses do not affect the material outside the ablation hole. Multiphoton absorption allows one to ablate even with photon energies below the indirect band gap of 5.4 eV (for example with 248 nm pulses) [63]. The same effect enables one to structure quartz with femtosecond pulses in the near infrared range (790 nm) [64]. Selffocusing phenomena even allow for a three-dimensional microstructuring [65], which might be used for the generation of photonic elements. An overview of the achievable structural dimensions can be found in [66]. In Figure 3.4 minimum structure sizes are demonstrated that can be achieved by simple projection of 193 nm UV light. More advanced methods such as phase-controlled multiple beam interferences result in nanoscaled surface structures [67,68].

It should be noted, however, that laser ablation with the help of femtosecond pulses is not a priori first choice for clean structure formation in the submicrometer range. The light intensity is so high that color centers are formed in dielectrica which lead to bulk absorption processes. If one irradiates the sample with more than one pulse, an explosion-like material evaporation follows an incubation phase. Time-resolved measurements of 'laser damage' of dielectrics via visible laser light (526 nm and 1053 nm) suggest that plasma formation is responsible for the de-

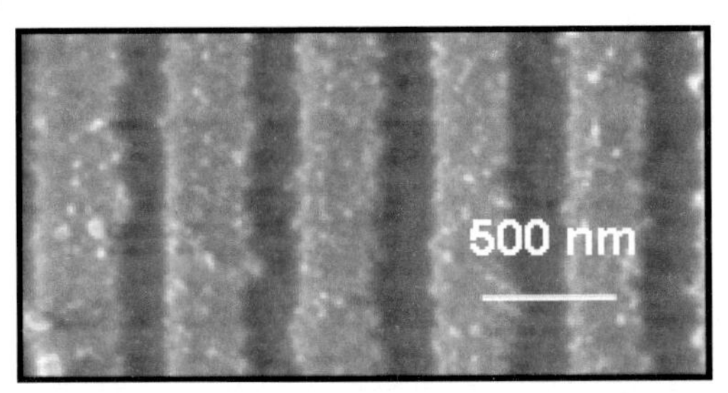

Figure 3.4 Minimum structure size achievable with single shot excimer laser ablation (193 nm) of ultrathin gold films on glass. R. Frese, NanoSYD, private communication 2008.

struction with pulse durations below 10 ps. Laser pulse durations above 100 ps destroy materials via melting and evaporation processes [69].

In order to write microstructures in ceramics such as aluminum oxide (Al_2O_3), which are difficult to manufacture with conventional methods due to their hardness, brittleness and chemical, electrical and thermal resistivity, nanosecond pulses are sufficient [70]. An interesting application of microstructured insulator surfaces are stepped dielectric transmission masks, which in a subsequent step are used to generate via laser ablation three-dimensional structures in other polymers or dielectrics [71]. As an example Figure 3.5 shows a Fresnel lens with micrometer dimensions, which has been generated by a single excimer pulse using a laser-structured mirror as mask. Three-dimensional structures can be induced via spatial varaiation of the reflection coefficient inside the mirror mask.

Laser Photopolymerisation

Micro-dimensioned three-dimensional structures have been generated within the last decade often via photostimulated reactions inside the diffraction-limited focus of a laser [72, 73]. Since the reactions need a threshold photon-density for activation, which can only be achieved within the maximum of the exponentially decreasing Gaussian profile of the laser focus, the minimum structure size is given by the size of the beam waist. In fact, using this method structure sizes below 100 μm are seldom achieved.

In order to achieve structure sizes below the Rayleigh limit multiphoton processes might be applied [75]. A special class of resins[4] can be polymerized only via two-photon absorption. Polymerization results in a glass-like material with index of refraction $n = 1.56$; thus the result-

4) For example, SCR500 from JSR,Japan or Norland NOA63.

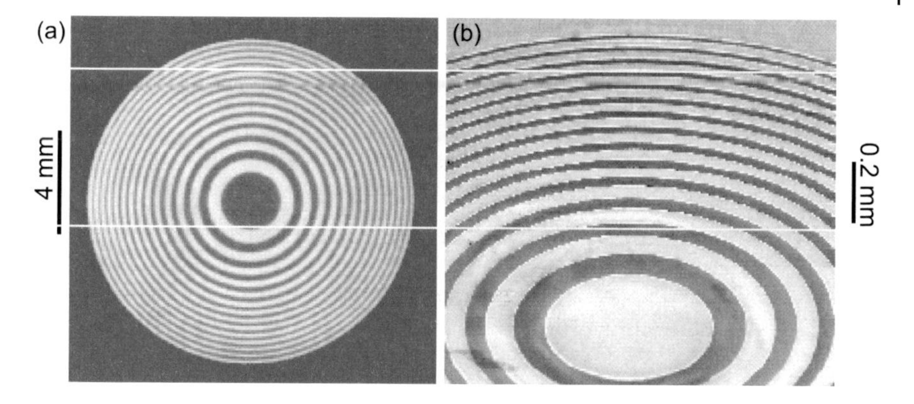

Figure 3.5 (a) Microscopy image of a zoneplate, written with an 193 nm excimer laser into a dielectric mirror, which is highly reflecting at 248 nm. (b) SEM image of a micron sized zoneplate, which has been written with the help of the mirror described in a) and a single 248 nm excimer laser pulse into a thin gold film on a dielectric substrate. [74]

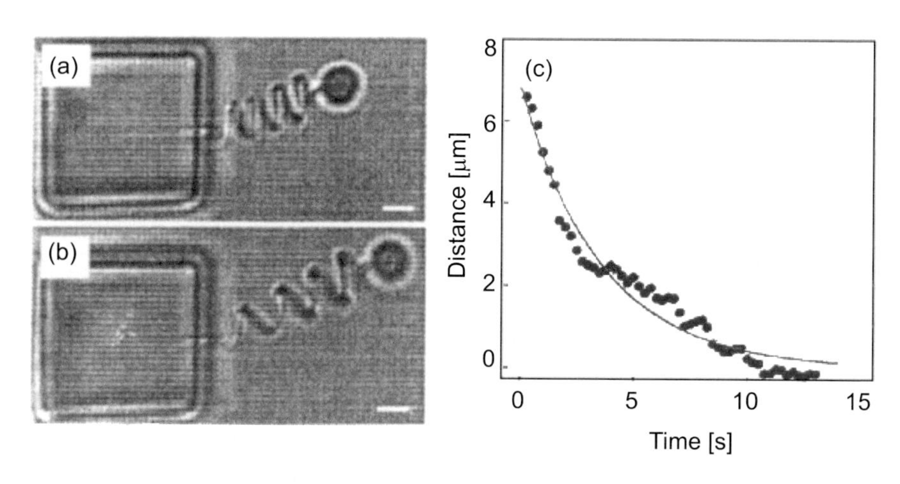

Figure 3.6 Micro oscillator in ethanol, generated via femtosecond two-photon photopolymerisation. The white line corresponds to 2 μm; the fabrication precision is 150 nm. (a) Equilibrium-, (b) expansion-state of the oscillator. (c) Retraction time as a function of expansion of the spring. The expansion has been induced by capturing the styrol sphere in the focus of a laser and drawing it with the laser focus. Reprinted with permission from [76]. Copyright 2001 Nature.

ing structures can be manipulated within the laser focus with the help of photon pressure. In Figure 3.6 vibrational motion is induced, in Figure 6.45 rotational motion. Structures are defined with a precision between 150 nm [76] and 500 nm [77].

3.1.2
Nanoimprint Methods

Besides generation of nanoscaled structures the precision with which they can be replicated is of huge importance for commercial applications. A standard method is the fabrication of a mask via lithographic techniques, which is subsequently multiplied via melting in thermoplastic materials (e.g. PMMA). For the fabrication of compact discs, structure sizes below 1 μm are obtained, for digital versatile discs (DVD) even below 400 nm with depths of the order of 100 nm.

In principle one should be able to imprint structures into polymers down to characteristic sizes of 10 nm, which corresponds to the volume that the polymer chain occupies (nanoimprint or 'soft lithography' [78]). Such techniques exist since the mid-1990s and one can replicate substrates up to several square centimeters with structure sizes of a few hundred nanometers [79–81] or even below [82]. Aspect ratios (depth to width) of up to 6:1 can be obtained.

The main problems of this method are inhomogeneous sticking and missing material transport; thus, high temperatures and high pressure in general improve the printing quality. Optimal conditions are around 100 bar hydraulic pressure and temperatures around 90 K above the glass temperature of the used polymer (in the case of the polymer Plex 6792 about 350 K) [83]. Systematic investigations have shown that periodic patterns and small scale structures (100 nm structure width) are nanoimprinted best with a negative stamping pattern [83].

3.1.3
Nanostructures via Scanning Probe Methods

Scanning probe microscopes are useful not only for imaging of surfaces with atom resolution but also for manipulation on an atom scale and for spectroscopy. This manipulation works under clean conditions in ultrahigh vacuum but also under ambient air or in the aqueous phase. Hence the conditions for the generation of biomolecular nanoarchitectures are fulfilled.

Besides tip-induced metal deposition [84] and local oxidation of various surfaces (silicon, graphite and metal surfaces) the scanning force microscope enables one to form structures in organic thin films mechanically [85, 86], thermally [87] and field-induced [88]. Even chemical reactions can be locally catalyzed [89]. A biologically relevant application is the deformation of a lipid double layer (membrane) with a force microscope tip – at this deformation point an enzyme from the solution can

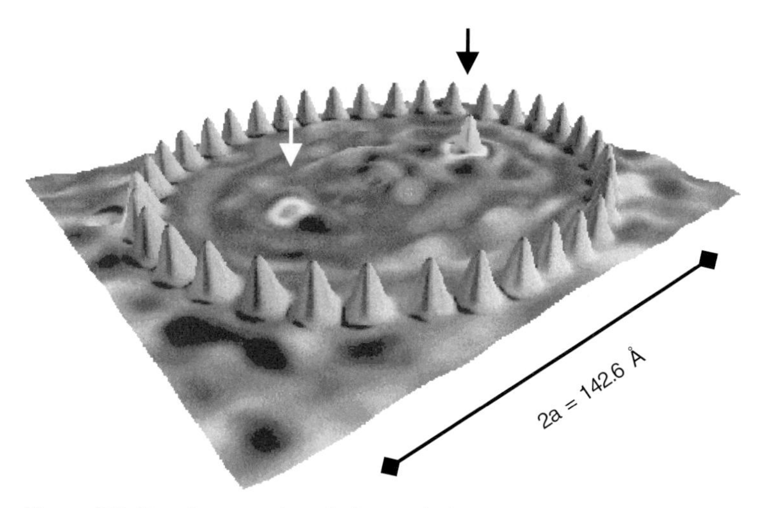

Figure 3.7 Quantum corral made from cobalt atoms on a Cu(111) surface at T=4 K. The large semiaxis a of the ideal ellipse is also plotted. The black arrow denotes a magnetic cobalt atom inside of the quantum corral, the white arrow its 'ghost image'. Reprinted with permission from [95]. Copyright 2000 Nature.

attack to cleave a fatty acid [90]. By this process the membrane can be modified with arbitrary, nanometer-sized structures.

Both the generation of nano scaled wrinkles as well as the transfer of atoms from the tunneling tip to the surface can be optimized by irradiating the tip with laser light [91,92]. This effect is due to electromagnetic field enhancement in the near field of the tip ('FOLANT': focusing of laser radiation in the near field of a tip). It allows one to obtain a lateral resolution of about 10 nm.

An impressive example for manipulation with the scanning tunneling microscope (STM) on the *atom level* are 'quantum corrals' , which have been fabricated at very low temperatures (4 K, to avoid thermal smear-out of the effects) on metal surfaces (Figure 3.7) [93]. The individual posts of the corral fence are represented by iron or cobalt atoms, which have been moved over the surface by narrow contact with the tip of the scanning tunneling microscope [94]. The wave pattern (i.e. energy-resolved Friedel oscillations) inside and outside the corral is induced by quantum mechanical interference phenomena, which result from scattering of surface electrons with the artificial adatoms.

The amplitude of the interferences is very small, typically a few percent of the height of an atom on the surface. Such shallow amplitudes can be detected only at very low temperatures (i.e. for minimum deterioration by thermal movements of the crystal and maximum resolution of the STM) and only for surfaces with nearly ideal two-dimensional electron gas. If these conditions are not fulfilled, the projection of the bulk states on the surface would dominate the STM movement. Such two-dimensional electron gases exist on the densely packed (111) surfaces of rare metals as delocalized 'Shockley' surface states.

If one uses the spectroscopic possibilities of the STM, individual magnetic atoms can be imaged using the Kondo effect [96]. This effect is observed, if a well localized magnetic perturbing unit (e.g. an individual magnetic atom) exists in a non-magnetic environment. At sufficiently low temperatures the spins of the conduction electrons around the magnetic atom are aligned in order to shield the spin of that atom – similar to the charges in the neighborhood of a conduction surface, which are shielded by the movements of the conduction electrons and a resulting counter field. This 'many-body spin state' becomes spectroscopically observable as a resonance in the density of states of the surface. The density of states $\rho(r, e_F)$ in turn is imaged by the STM.

If one places a magnetic atom in one of the focal points of the elliptical quantum corral, mirror images are generated at the second focus due to the electronic Eigenmodes of the corral (Figure 3.7). The projection medium in this case are the two-dimensional electron waves in the Shockley state of the copper surface. The mirror images have the same spectroscopic signatures as the originals, that is, the electronic structure of the real cobalt atom has been projected exactly into the second focus of the ellipse.

The whole exercise represents electron optics on surfaces on an atom scale. Such optics could be used for future fabrication of atomic information technology elements. The advantage compared to conventional electron transport along the surface by wires is that the wave nature of the electron is employed, meaning that one could also use all the other advantages known from conventional optics, such as destruction free penetration, generation of solitons, holographic aspects and so on.

3.2
Bottom-up Methods

In order to obtain well defined microscopic growth of nanostructures without the need for subsequent manipulation of the structures on the nanometer scale, one needs a precise knowledge of the crystallographic parameters of the substrate and the adsorbate. Further, growth energetics and kinetics have to be well understood [97]. This is of course also an important prerequisite for the generation of any crystallographically well-oriented ultrathin film. For the fabrication of laterally and vertically nanostructured films without using templates the requirements on the microscopic knowledge are even higher, especially if defined layer-by-layer growth or the generation of nanoscaled lateral structures are a part of the growth recipe.

3.2.1 Epitaxial Growth

Usually one defines epitaxial evaporation as the sputtering or evaporation of one material on another or the same material with the goal of generating a new crystalline structure which can be generated only with difficulty from the bulk crystal of the material [98]. Among epitaxial techniques the term 'hetero-epitaxy' denotes the application of different adsorbate and substrate materials and 'homo-epitaxy' the same adsorbate and substrate materials.

The adsorbate does not necessarily grow as a well-ordered film ('layer-by-layer' Frank–Van der Merwe growth [99]) but can also form three-dimensional aggregates ('clusters') (Volmer–Weber growth [100]). The former growth mode is often used for the growth of metals on metals, the latter more often for the growth of metals on dielectrics. Which mode is chosen is determined by structural points (misfit between lattice constants of substrate and adsorbate) and by energetic considerations. These are, for example, the binding energies between the atoms of the adsorbate, the atoms of the substrate and between adsorbate and substrate. Cluster growth is apparently preferred if adsorbate-adsorbate binding energies are larger than adsorbate-substrate binding energies. During growth additional energies start playing a role, for example, strain within the growing layers. In such a case a two-dimensional film might first grow, followed by cluster growth on top (Stranski–Krastanoff growth [101]). This kind of self organization, which can lead to clusters with very narrow width distributions, is discussed in more details later. Of course, the *de facto* observed growth mode depends not only on the binding energies but also on the growth conditions (deposition rate, substrate temperature) [102] .

If one assumes for sake of simplicity that the islands on the surface possess the shape of droplets, it becomes possible to quantitatively discriminate the different modes by considering the contact angle ϕ between the edge of the droplet and the surface plane. The equality of forces (Figure 3.8b) results in the Young–Dupre equation

$$\gamma_{SV} = \gamma_{DS} + \gamma_{DV} \cdot \cos\phi, \tag{3.3}$$

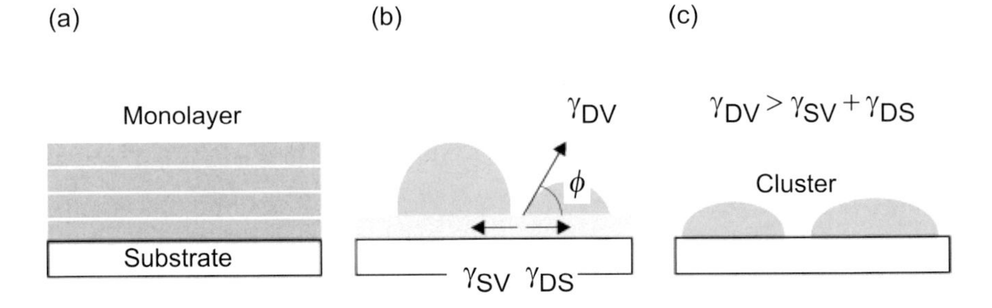

Figure 3.8 Growth modes: (a) Frank–Van der Merwe, (b) Stranski–Krastanoff, (c) Volmer–Weber.

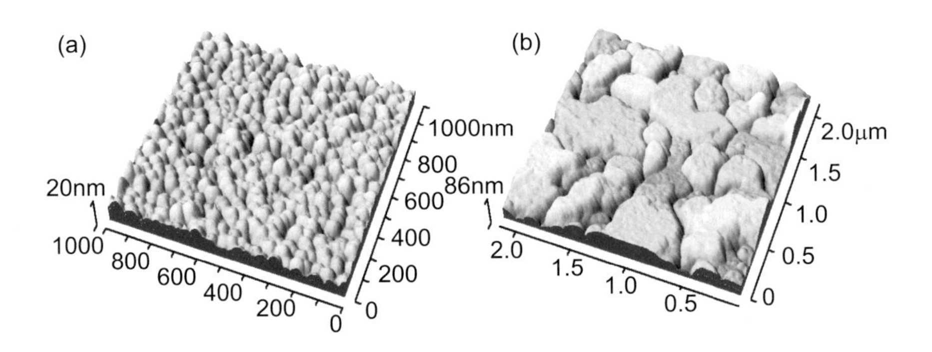

Figure 3.9 Atomic force microscopy images of thin gold films on mica: thickness 10 nm (a) and 50 nm (b). Initially small aggregates are formed, which form with increasing film thickness a continuous film of large, flat islands.

with γ_{DV} the surface tension between droplet and vacuum, γ_{SV} the surface tension between substrate and vacuum and γ_{DS} the surface tension for the interface between droplet and substrate. For $\phi \to 0$ a two-dimensional layer will grow on the surface, while for $\phi > 0$ droplets grow (Figure 3.8c). A Stranski–Krastanoff growth results if the increasing strain energy during film growth is initially used to relax the adsorbate lattice until its lattice constant agrees with the lattice constant of the substrate lattice. Following the growth of a monolayer, this energy is no longer necessary for lattice relaxation. However, since this energy is further increasing within the adsorbate layer, droplets are formed on the first monolayer [103]. Extended numerical simulations and scanning tunneling microscopy experiments have led to a very good understanding of the initial states of epitaxial metal [104] as well as three-dimensional island growth [105].

Figure 3.9 shows gold films of different thickness on a mica substrate, which have been generated via epitaxial growth. As can be seen, the roughness of the surface changes as a function of film thickness. A convenient way to describe the roughness is to fit a Gaussian function

$$G(\vec{r},\vec{r}') = \delta^2 exp(-\frac{|\vec{r}-\vec{r}'|^2}{\lambda_0^2}) \quad (3.4)$$

to measured height distributions. This function represents the height-height correlation between different points of the surface profile S($\vec{r}$) (see also [106] and [107] for further discussions). In Equation 3.4 $\vec{r} = (x,y)$ denotes the position on the surface, λ_0 the transversal correlation length and δ the average height, $\delta = \sqrt{\bar{S^2}}$. For Figure 3.9 one finds $\delta = 2.53$ nm and $\lambda_0 = 25$ nm for the 5 nm thick film and a larger roughness $\delta = 13.65$ nm and $\lambda_0 = 115$ nm for the 50 nm thick film.

Rough metal films of this kind are of great interest for their possible optical applications since they consist of strongly polarizable particles, the optical properties of which are determined by morphology and electronic structure. Due to the weak binding energy between metal atoms and dielectric (about 100 meV) the atoms have a high mobility on the surface. They will migrate over the surface and will be localized on defects such as edges, corners or dislocations as well as already adsorbed atoms. At these places they will form aggregates (binding energies between metal atoms of the order of electron Volts), which eventually coagulate. The roughness of the thin film that is generated in this way depends on the exact preparation conditions (surface temperature, defect density, flux of adsorbing atoms, kinetic energies etc.). Despite the fact that this type of metal film growth has been investigated for a considerable length of time, for all with alkali metals [108] and for ultra thin gold films [109–111], the final roughness can still only be approximated.

Growth via thermal evaporation is a statistical process and thus the fabricated structures will possess only a small degree of regularity. The resulting islands have a wide size distribution [112]. Optimized methods of molecular beam epitaxy (e.g. growth interruption) result in much narrower size distributions [113]. Other approaches such as laser desorption [114] or the inclusion of selforganization phenomena into the epitaxial growth process also result in more homogeneous morphologic distributions of islands on surfaces.

3.2.2
Self Organisation

Films

Future optoelectronics on the basis of molecular or nanostructured architecture will most probably include basic elements made from organic molecules. The main advantage of organic molecules is their flexibility, that is, the possibility to modify their optoelectronic and structural molecular properties by synthetic chemistry and the ability to combine them with other materials such as polymers. The goals of such new optoelectronics range from the fabrication of ultrafast optical biosensors [115] and lithographic applications [116] to micro- and nanostructured optical frequency doublers on the basis of waveguides [117]. It is therefore important to have possibilities at hand to fabricate monomolecular organic films with well defined crystalline structure [118].

Various light emitting organic molecules have been investigated via organic molecular beam epitaxy (OMBE) growth in terms of their ability to form well-organized, ultrathin organic films. These include, thiophenes [119], PTCDA [120], pentacene [121], para-phenylenes [122] and anthraquinone [123]. These are all molecules similar to the phenylenes shown in Figure 3.16.

Besides OMBE, methods based on self organization are also appropriate growth tools[5].

Since ancient times it is known that oil spread on water forms an ultrathin film, which reduces the surface tension of water. Aristoteles used a few liters of oil to smoothen the waves over a whole harbor area. In the nineteenth and twentieth centuries Agnes Pockels [124] and later Katharine Blodgett and Irving Langmuir [125] demonstrated that the use of a Teflon barrier drawn over the surface of fatty acid coated water results in a monomolecular film of high quality (Langmuir–Blodgett method) [126]. The thickness of this LB film ('Langmuir–Blodgett' film) depends on the number of 'basic groups' (CH_2 groups in the case of a hydrocarbon fatty acid film) and the tilt angle to the substrate. For cadmium arachidic acid ($[CH_3\text{-}(CH_2)_{18}\text{-}COO^-]_2Cd^{++}$) the thickness of a monolayer is 26.4 ± 0.1 Å [127].

As shown schematically in Figure 3.10 the main driving force for the generation of a monomolecular film is the strong binding between the hydrophilic end groups of the fatty acid and the water surface. The hy-

5) Nature uses self organization in an infinite number of variations. This was realized a long time ago. For example, it has been known since 1925 that biological membranes consist of self organized double layers of amphiphilic molecules.

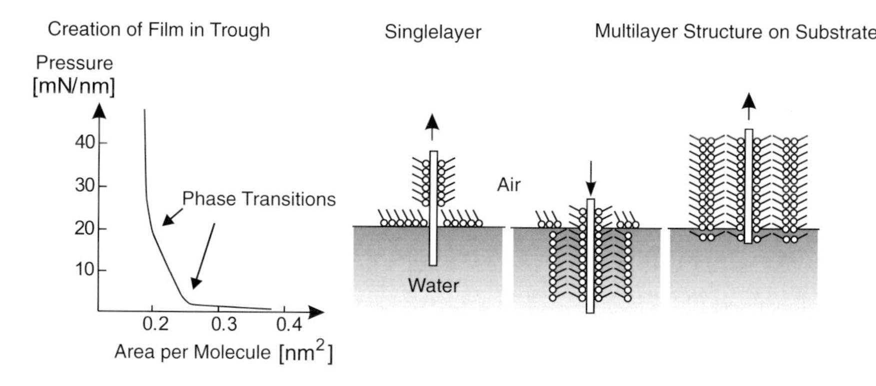

Figure 3.10 Generation of monomolecular films with the help of the Langmuir–Blodgett method.

drophobic end groups remain directed away from the water during compression with the teflon barrier. The mutual van der Waals interactions of the molecular chains lead to the expected film alignment. The order of the film can be determined indirectly from the increase in surface pressure: an ordered monomolecular film can be compressed similar to a fluid only weakly due to the repulsive interaction of the molecules.

The film from organic molecules is transferred onto the target substrate by drawing the substrate perpendicularly through the fatty acid film on the water surface. If the substrate surface has an appropriate polarity the hydrophilic end groups of the fatty acid are detached from the water surface and stick on the substrate surface. Useful substrates are insulators (e.g. a microscope slide) or metals (e.g. thin gold films). In order to deposit further films, the substrate is dipped and drawn again, keeping the surface pressure constant by moving the teflon barrier. Using this method even in the first publications of K.Blodgett multilayers up to a few thousand layers (some μm) could be reported [125,128].

Instead of taking advantage of the polarity of the organic molecules or of their hydrophobic nature one can also use chemical reactions between one of the end groups of the molecule and the substrate surface to facilitate defined film growth. The sulfuric end group of alkane thiols such as docosan thiole, CH_3–$(CH_2)_{21}$–SH, C_{22}, chemisorbs with a binding energy $E_b \approx 2\,\mathrm{eV}$ [129] on gold (cf. Figure 3.14 for a graphical representation of the alkane thiols). This strong chemisorption energy builds the basis of a self-organization process that leads to well organized monomolecular films of alkane thiols on gold surfaces. The sulfuric groups represent the anchors on the surface and the CH_3 groups point away from the surface [130]. Films of this kind are usually called SAMs ('self assembled monolayers') [131]. The fraction of uncovered surface for a SAM film is much smaller than the approximate 30%, which is typical for an LB-film.

In the twenty five years since the first description of the method [132, 133] a multitude of applications have been investigated. In contrast to the LB method, which asks for a relatively large technical effort (a 'Langmuir–Blodgett' trough), as well as experience in using chemicals and very careful cleaning procedures, monomolecular SAM films can be produced by simply leaving the gold substrates a few hours in an roughly one milli molar ethanol solution of the respective alkane thiol. However, due to the strong gold-sulfur interaction and the weak interaction with the methyl end group it is difficult (though possible [134]) to adsorb more than one monolayer onto the substrate. That means that the thickness of the organic layer can be varied only by variation of the length of the hydrocarbon chain (up to a length of 3 to 4 nm), via 'self replication of amphiphilic monolayers'[6] or via attachment of a special functional end group in place of the inert methyl end group onto which further layers can be attached [136].

In most cases, however, the change of the end group results in a large disorder in the monomolecular films. As an example Figure 3.11 shows models of an OH-group terminated and an COOH-group terminated alkane thiol film. The models result from a detailed surface analysis using spectroscopic methods (NEXAFS).

The most important advantage of using a self organized monomolecular organic film inside a complex structure is that a structural manipulation and characterization on a microscopic level becomes possible. This of course relies on a characterization of interface quality with microscopic surface-analysis methods.

Structural information such as crystallinity or tilt angle of the individual molecules with respect to the surface plane become available via X-ray diffraction [138–140], electron microscopy [141], metastable spectroscopy (MIES) [142], electron-(LEED) [143] or atom beam diffraction [144]. Complementary information in real space is provided by atomic force microscopy [145] or scanning tunneling microscopy [146, 147]. Optical methods allow thickness measurements (ellipsometry)[126], information about internal motions (infrared and Raman vibrational spectroscopy) [148] or concerning the growth dynamics (optical frequency doubling [149]). A review article on structure and growth of SAMs can be found in [150].

6) Self replication is a variant of the LB method, which uses double layers of self organizing silanes, which are polarized in solution with hydrogen allowing the next double layer to grow [135].

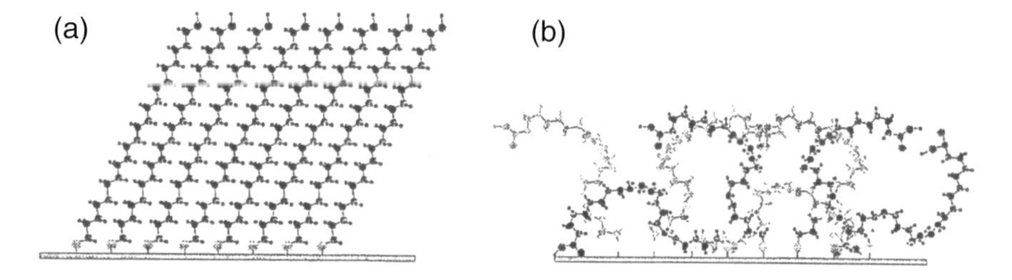

Figure 3.11 Monolayers of long chain alkane thiols, terminated with OH (a) and COOH (b). Reprinted with permission from [137]. Copyright 1997 Elsevier Science.

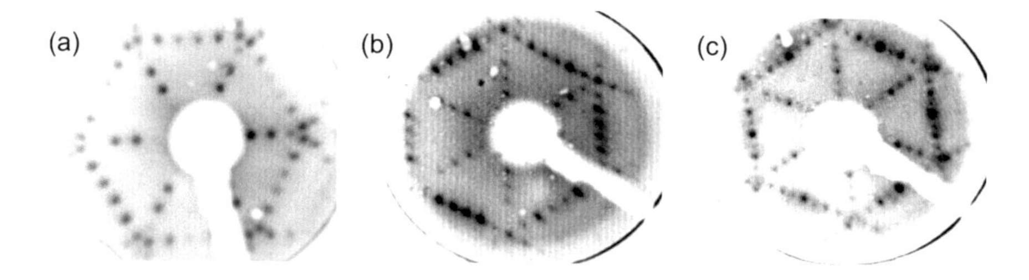

Figure 3.12 (a) LEED-pictures of a clean gold surface at $E_{el} = 120\,\text{eV}$, (b) a heptane thiol covered gold surface at $E_{el} = 49\,\text{eV}$ and (c) a decane thiol covered gold surface, also at $E_{el} = 49\,\text{eV}$.

An example of the information that can be obtained using LEED is shown in Figure 3.12. Diffraction spots from a plain, single crystalline gold surface (a) and a gold surface, covered with an ultrathin film of heptane thiol (b) and decane thiol (c) are visible. From a comparison of these measured with calculated diffraction spots and assuming a certain structure of the overlayer on gold it follows that the organic molecules at the low coverage used here are oriented parallel to the surface. The structure corresponds to the structure shown in Figure 3.13b, which in turn agrees with atomic force microscopy images (Figure 3.13a).

If the molecules are oriented parallel to the surface, the size of the unit cell of the growing crystal phase increases with increasing chain length, which results in a characteristic variation of the LEED images (Figure 3.12b and 3.12c). From number and position of the diffraction spots one can deduce size and orientation of the unit cell of the single crystalline growing organic thin films. With increasing surface coverage the orientation of the organic molecules on the surface changes; namely, they tilt upwards. The corresponding electron diffraction images show that behavior clearly [143]. Low energy electron diffraction is thus a use-

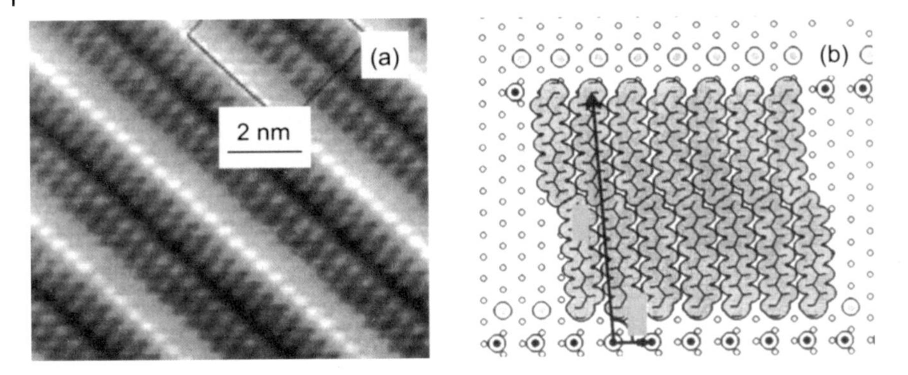

Figure 3.13 (a) Atomic force microscopy of decane thiol molecules, laying on Au(111). (b) Real space structure of the stripe phase, shown in (a). Reprinted with permission from [151]. Copyright 1998 American Chemical Society.

ful way to control the single crystalline growth of the organic films. Such a control is important for possible applications of the films as ultrathin dielectric spacer layers between metals [152] (Figure 3.14), as gapless coating around reactive materials or as a sliding layer between nanoscaled mechanical elements. Possible electronic applications and corresponding problems are discussed in more details in Section 6.2.2.

Tubes and Wires

Under certain growth conditions strongly discontinuous thin film growth of organic molecules on specific dielectric surfaces is observed. The resulting needle-like structures with heights and widths of several tens to several hundred nanometers, and lengths ranging from micrometers to millimeters are similar to those formed via catalyst or template-wetting [153] induced growth normal to the surface plane (cf. Figure 6.1). They also resemble needle-shaped organic catalysts formed by self-assembly in liquids [154–158].

The obvious challenges for the latter techniques are the controlled growth of crystallites of predefined shapes and predefined mutual orientations and their transfer onto more complicated target substrates. The huge advantages, however, are large design flexibility, excellent device integrability, and potentially a much improved performance of resulting devices [159,160], similar to the evolution from liquid-crystal-based flat screens to organic light-emitting-device based screens.

Hence such nanotubes (hollow), nanowires (filled) or 'nanofibers' are an attractive crystal modification [161]. In many cases they are formed

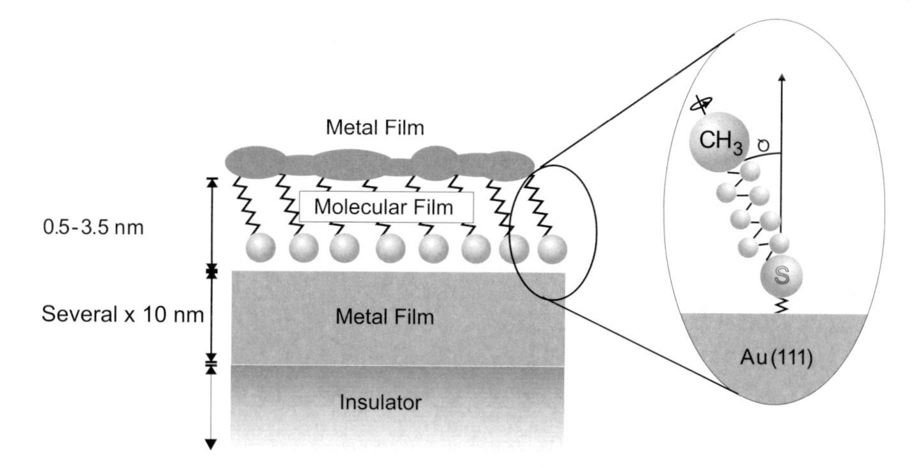

Figure 3.14 Monomolecular organic thin films can be used as insulating spacer layers, for example, between a thin, on an insulator deposited metal film and a further metal film, deposited onto the organic film. The 'enlarged' view shows a side view of alkane thiols that grow on Au(111) at a tilt angle of 30° with respect to the surface normal.

from inorganic compounds such as silica, ZnO, GaN, InP, and so on [162–166].

In order to resolve the above listed challenges, surface-based growth and the use of modified *organic* molecules as building blocks might prove a more direct approach to molecular nanotechnology. Needle-shaped growth of organic molecules such as para-phenylenes has already been observed on specific dielectric surfaces, namely muscovite mica surfaces [167, 168]. A closer look, for example, with an epifluorescence microscope, shows that the needle-like structures are mutually parallel oriented (Figure 3.15c). Between the parallel oriented needles smaller aggregates are found, which turn out to be the precursors of the needles. The area immediately around the needles is free from the small aggregates.

The 'needles' are made of hexaphenyl molecules, which are flat laying and oriented nearly perpendicularly with respect to the long needle axis. Detailed investigations using electron- and X-ray diffraction reveal the unit cells of the crystallites and thus a precise microscopic image of the structure of the needles (Figure 3.17b). Growth of needle-like crystalline aggregates is also found on other substrates such as alkali halides. Height and width of the needles are similar. However, the needles grown on muscovite mica are much longer (up to several hundred microme-

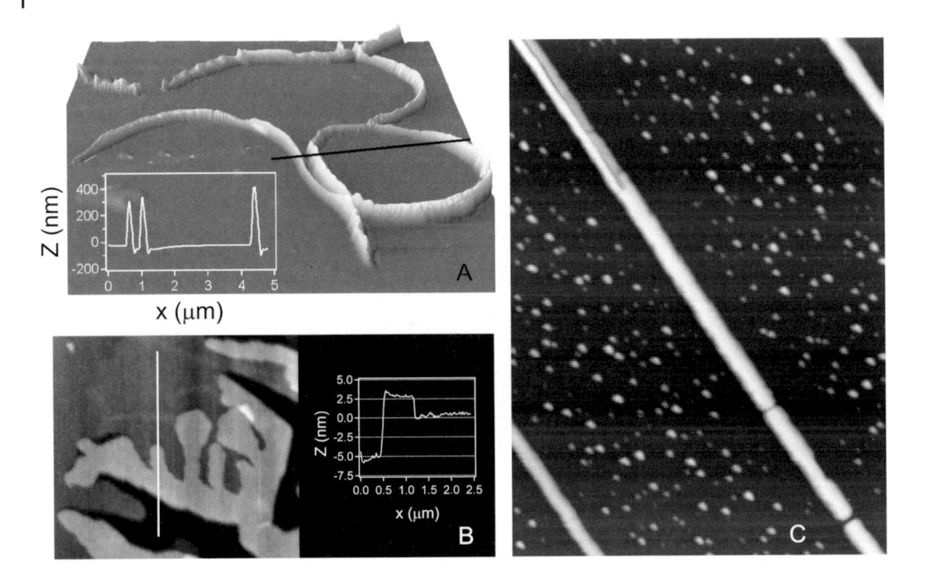

Figure 3.15 Force microscopy images of hexaphenyl aggregates, deposited on non-polar (a) and polar (c) mica surfaces. Layers of standing molecules are formed (b) as well as ring-shaped (a) or needle-like (c) structures, which consist of laying molecules.

ters) and they are very well mutually oriented. Detailed investigations via electron diffraction as well as optical methods have shown that the growth mechanism of the needles is influenced by strong electric surface dipole fields that are induced on the mica surface upon cleavage. These dipole fields possess two possible orientations on a cleavage plane of mica (three, if one takes into account lower lying planes), and they do not exist on alkali halide surfaces. The dipole fields induce a dipole moment in the polarizable organic molecules, leading to an attraction via dipole-induced dipole forces and thus to an alignment of the individual organic molecules along the surface dipole orientations. Subsequent molecules grow side-by-side on the adsorbed molecules on the mica surface, if they possess enough surface mobility (i.e. if the surface is warm enough), leading to the generation of aligned, needle-like aggregates with very well defined molecular orientations.

The oriented growth process includes surface dipoles on the mica surface as an important ingredient and it has thus been called 'DASA', dipole-assisted self assembly [167].

Figure 3.15 shows several growth modes of organic molecules on mica surfaces.

In Figure 3.15a hexaphenylene molecules have been deposited on a mica surface which has been treated with water prior to organic molecule deposition. Ring-like aggregates are formed, which emit intense blue light after UV excitation. Below the aggregates a continuous film of molecules has been formed (Figure 3.15b). Optical analysis reveals that the continuous film consists of upright oriented molecules, while the aggregates are formed from molecules that are oriented parallel to the surface plane. This observation agrees with AFM data, which show that the continuous film consists of stacked planes, the step height of which correspond to the length of hexaphenylene molecules (see also the cross section in Figure 3.15b).

As indicated above, nanoscaled structures or ultrathin films made of phenylene oligomers are of huge interest for future display technology on the basis of organic materials since the molecules show strongly polarized electroluminescence and blue light emission after UV excitation. Both absorption- and emission wavelengths can be varied by changing the number of phenylene rings (Figure 3.16).

A possible application in nanoscaled waveguides is shown in Figure 3.17. By fast cooling, a break in the needle of a few ten nanometers width has been generated (force microscopy image to the right-hand-side), which does not influence the orientation of the needle, but scatters

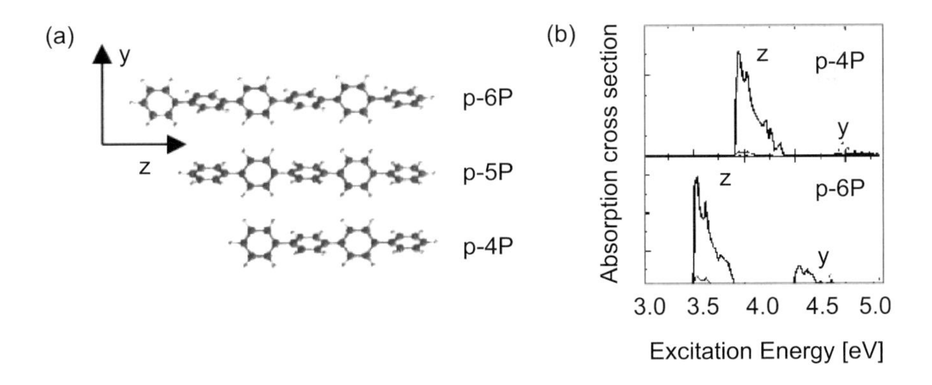

Figure 3.16 (a) Molecular structure of quarter phenyl (p-4P), penta phenyl (p5P) and hexa phenyl (p6P) oligomers. (b) Calculated absorption spectra along the long axis (z) and perpendicular to it (y) [169]. A strong anisotropy of the absorption cross section is observed. With increasing molecule length the resonances are shifted into the red spectral range (lower transition energy).

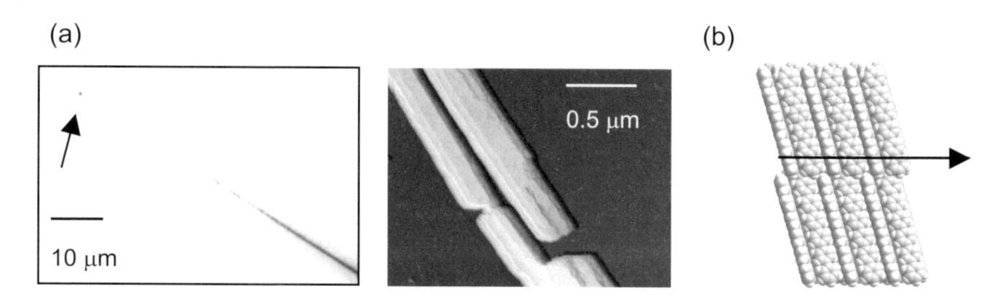

Figure 3.17 (a) Fluorescence microscopy of a single needle made from organic molecules, which is excited locally on the right-hand-side with UV light. The light is guided through the needles and scattered on the left-hand-side (arrow). The detailed AFM image shows that the scattering occurs on a nanoscaled break of the needle. (b) The individual molecules are oriented approximately perpendicularly to the long needle axis.

light [170]. The guided light has been induced inside the waveguide by UV excitation, and it is detected following its propagation through the nanowire via scattering at the break.

The needle growth process is thus dictated by the strength and orientation of the dipole fields on the surface, the polarizability of the molecules and their mobility on the surface. It works especially well on muscovite mica due to a favorable quasi-epitaxial relationship between adsorbate and substrate. Domains with specific dipole directions can be huge on mica (of the order of square millimetres to centimetres) and consequently huge domains with parallel oriented nanoaggregates can be formed. The temperature window within which the surface has to be kept for the successful growth of long needles is only of the order of 20 to 30 K; at lower temperatures quasi-continuous films consisting of very short, dense needles are formed; at higher temperatures desorption takes place. This strong temperature dependence allows one to grow the needles at predefined spots on the surface via, for example, local laser heating [171], and enables control over the environment of individual needles.

The fact that para-phenylenes plus muscovite mica constitute an unique combination from a crystallographic and growth dynamic point of view has resulted not only in unique nanoaggregates but obviously also limits the potential range of applications of these 'nanofibers'. However, two recent developments have opened the door to a much wider application potential: i) the possibility to transfer the nanofibers from the original growth substrate to any other substrate or into liquids [172]; ii) the possibility to change the functionality of a para-quaterphenylene block by synthesizing new molecules including para-quaterphenylenes with specific end-groups as well as the generation of aligned nanofibers from these functionalized molecules [173]. In terms of implementation of nanoaggregates into working devices the former development, (i), has enabled electrical conductivity as well as mechanical deformation measurements [174] on single nanofibers, whereas development (ii) has resulted in the growth of tailored nanoscaled frequency doubling elements [175]. Further device development is only a question of time [176].

Dots

In the previous two sections we have exemplified 2D film and 1D wire growth by organic molecules. 0D growth of dots is more often found for metallic or inorganic semiconductor adsorbates; note, however, that semiconducting organic quantum dots can also be fabricated, see, for example, Figure 4.15. If one allows diffusion limited mass transfer (e.g. by storing the new materials in ambient air), the average radius a_0 of islands on surface grows with a characteristic time t_c as

$$a_0(t) = a_0(t=0)\left(1 + \frac{t}{t_c}\right)^{0.25} \tag{3.5}$$

(thermal driven 'Ostwald ripening' [177]). This ripening process results in an array of well separated 'self organized dots' (SODs). Repulsive interactions between growing semiconductor islands on metals result in lattice structures with atom dimensions [178]. For the generation of metal islands on metals one takes advantage of 'strain relief' processes during the growth process [179], which also result in self organized nanostructures [180]. During growth of an adsorbate layer (e.g. copper) on a substrate with a different lattice constant (e.g. Pt(111)) strain is induced, which results in the formation of periodic dislocations. At the point of these dislocations thermal adsorbed metal atoms (e.g. iron) grow in periodic island patterns with nearly monodisperse sizes. It has been shown that strain-induced interactions between surface defects represent a generic driving force for the generation of self organized nanostructures on metal surfaces [181].

Pyramids

During homoepitaxial growth (e.g. copper on copper [182, 183] or iron on iron [184]) regular pyramidal structures grow under appropriate conditions, with well defined distance between them and with well defined wall slopes (Figure 3.18). This process is induced by a growth instability due to a step barrier ('Ehrlich–Schwoebel barrier' [185,186]), which counteracts the downward diffusion of adsorbed atoms. As a consequence the growth of higher laying islands on lower laying islands is more fa-

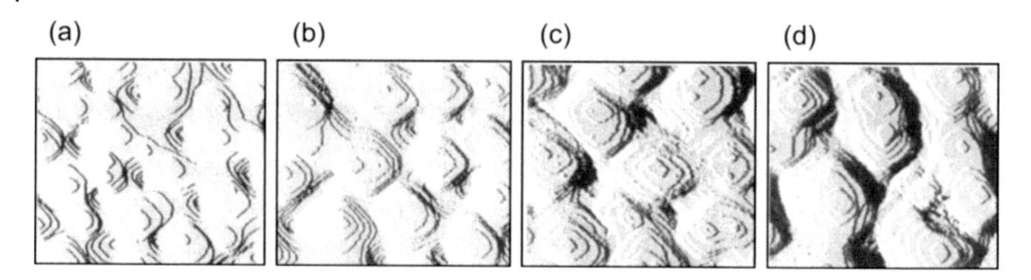

Figure 3.18 Atomic force microscopy of Cu islands on Cu(100) as a function of coverage: 21 monolayers (a), 42 monolayers (b), 73 monolayers (c) and 115 monolayers (d). Image size 100 x $100\,\text{nm}^2$. Reprinted with permission from [183]. Copyright 1997 The American Physical Society.

vorable. Copper pyramids on Cu(001) surfaces have been observed at surface temperatures near room temperature [183] but also at temperatures as low as 160 K [182]. The flux of incoming atoms was about one monolayer per minute and the energetic value of the Schwoebel barrier roughly 125 meV [187].

As seen in Figure 3.18, the wall slope of the pyramids increases with increasing coverage since the surface roughness and the mutual distance $L(t)$ of the pyramids grows with coverage time t as $L(t) \propto t^{0.25}$. Simultaneously the gaps between the pyramids are filled with smaller structures, which means that the mutual distance of pyramids is well defined only for the uppermost layers. Above 100 monolayers the step density remains constant. From the constant broadening of electron diffraction profiles one calculates an average terrace width l. Together with the atom step height h one obtains the constant pyramidal angle:

$$\phi_0 = tan^{-1}(d/l) \quad . \tag{3.6}$$

A typical value for Cu(100) at room temperature is $\phi_0 = 2.4°$ [183]. The pyramids are thus very flat. Growth at lower temperatures (160 K to 200 K) results in steeper pyramids: the vicinal side faces of the pyramids change from the crystallographic orientation (115) at 200 K to the crystallographic orientation (113) at 160 K [182]. See also Figure 4.43.

Chains

One-dimensional chains from atoms or molecules can be obtained with the help of templates, for example via the use of a regular pattern of surface defects (cf. Figure 6.25) or surface charges (cf. Figure 3.15). Chains of C_{60} molecules have been grown using the stripe phase of

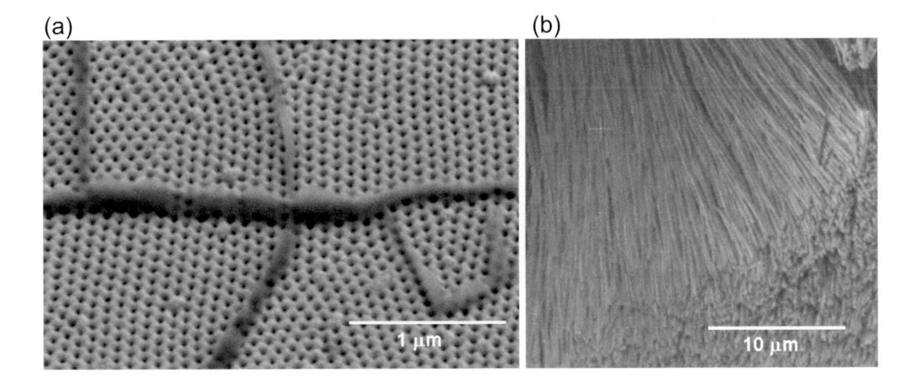

Figure 3.19 (a): SEM image of para-hexaphenylene nanofibers grown on an Au-coated alumina template. There appears to be a relationship between the pore-order and the growth direction of the nanofibers. G. Kartopu, M.Es-Souni, University of Applied Sciences, Kiel; M.Madsen, NanoSYD, Denmark, private communication, 2008. (b): Template assisted growth of crystalline organic tubes from molecularly assembled building blocks: tubes with an outer diameter of about 300 nm (defined by template geometry) after dissolving the inorganic template for taking the SEM-image. The wall thickness of the tubes depends on the growth conditions. The length is defined by the template thickness. Rods are obtained when using smaller template pores. M. Rastedt, K. Al-Shamery, University of Oldenburg, private communication, 2008.

alkane thiols on gold (Figure 3.13) as the template [188]. Even DNA is a well suited template for the generation of ultrathin metallic wires (cf. Section 6.4). One can avoid to use templates of that kind if one uses adsorbate-substrate combinations that result in an anisotropic, repulsive interaction between the adsorbate particles. An example is parallel rows of pentacene molecules on a Cu(110) surface [189]. The distance between the chains was as small as 3 nm.

Template Assisted Growth

Three-dimensional growth of nanoscaled aggregates can be directed by the lithographic formation of two-dimensionally, periodically ordered seed points for catalytic growth. Figure 6.1 shows a resulting distribution of lasing nanowires, which have been grown using such a method.

Another approach is to use periodically ordered arrays of defects or holes as growth templates, for example, macro- or mesoporous materials (Section 5.2 and Figure 5.3). It is seen that the periodic surface order, induced by the pores, influences both the growth along the surface (Figure 3.19a) as well as the growth perpendicular to the surface (Figure 3.19b). The former growth mode is induced by physical vapor de-

(a)

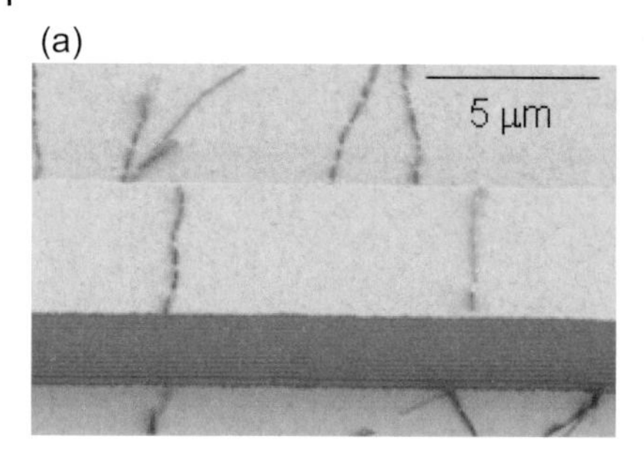

(b)

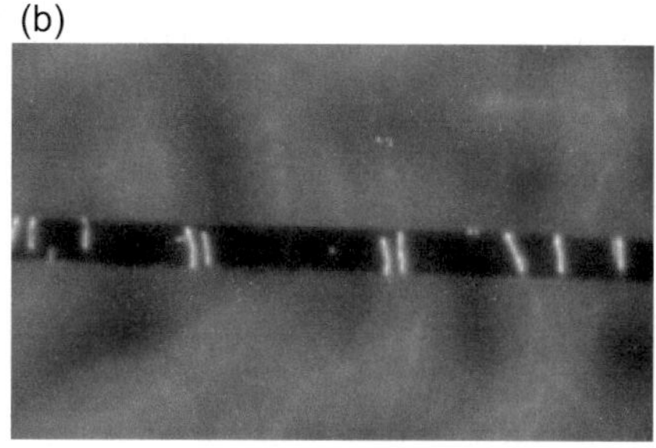

Figure 3.20 (a): SEM image of a lithographically fabricated ridge structures in silicon, coated with an ultrathin gold film. Nanofibers have been grown on top of it. The nanofibers emit blue light after excitation with UV light. (b): Fluorescence microscopy image of organic nanofibers, grown in an oriented fashion on top of the ridges shown on the left-hand side. M.Madsen, NanoSYD, Denmark, private communication, 2008.

position, whereas the latter growth mode of crystalline rods inside the pores is usually associated with the liquid phase. The advantage of this approach is that it is cheap, simple and reliable and thus allows in principle mass production of nanotubes and nanorods.

Problematic with the defect-assisted growth of nanoaggregates on surfaces as detailed in Figure 3.19 is that the chemically etched pores possess no long range periodicity. Such periodicity is more readily obtained by lithographically generated defect structures. In order to demonstrate how such micro- nanostructure formation on surfaces can enable one to tailor the growth of nanoaggregates, micron-sized ridge structures have been lithographically etched into silicon (Figures 3.20 and 3.21). Organic molecular beam epitaxy then results in directed growth of nanorods, namely perpendicular to the ridge edges. In comparison, unstructured surfaces lead to non-directed growth of nanoaggregates (Figure 3.21).

3.2.3
Deposition of Preselected Cluster Matter

Clusters possess extraordinary optical, electronic, magnetic and chemical properties, for example, size-dependent catalytic activity. They preserve these properties partially even in solution or following adsorption on surfaces [190]. An example is rhodium clusters on aluminum-oxide, which are catalysts for thermal dissociation of CO. The catalytic activity has a maximum for cluster sizes from 500 to 1000 atoms, probably because of the very high step density in this size range [191].

In order to obtain a well defined network of size selected clusters on surfaces via gas phase deposition [192], a number of technological prob-

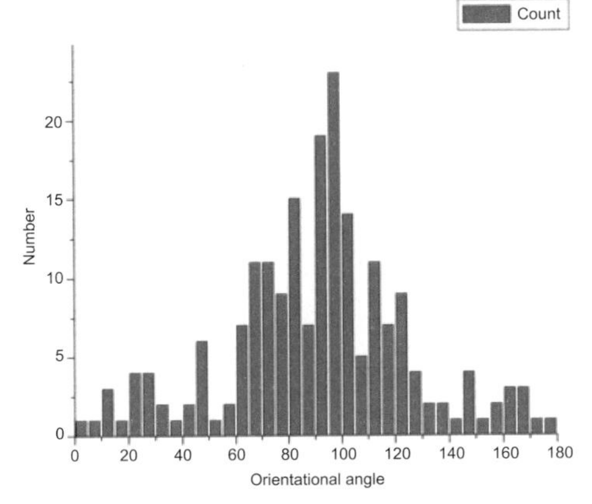

Figure 3.21 (a): Fluorescence microscopy image of growth of organic nanofibers on structureless (left and right lower corners) as well as microstructured (upper right corner) gold coated surface areas. (b): Measured distribution function of orientation angles of nanofibers on the structured part of the surface. A clear preference for growth perpendicular to the ridge edges is observed. M.Madsen, NanoSYD, Denmark, private communication, 2008.

lems have to be solved. A thorough overview of aspects of preparation of mass selected atom clusters in the size range 1 nm to 10 nm and their properties after 'landing' on the surface can be found in [193,194]. The influence of particle-surface interactions and particle-particle interactions are discussed, but also possible new electronic and magnetic properties that the nanoscaled systems on the surface might possess.

The shape and structure of the cluster on the surface depend very much on the initial size and on the energy of incidence. Molecular dynamic simulations for size selected silicon clusters impinging on graphite show that small clusters with small energies of incidence wet the surface nearly completely, that is form commensurate two-dimensional islands [196]. With increasing size the probability increases that the three-

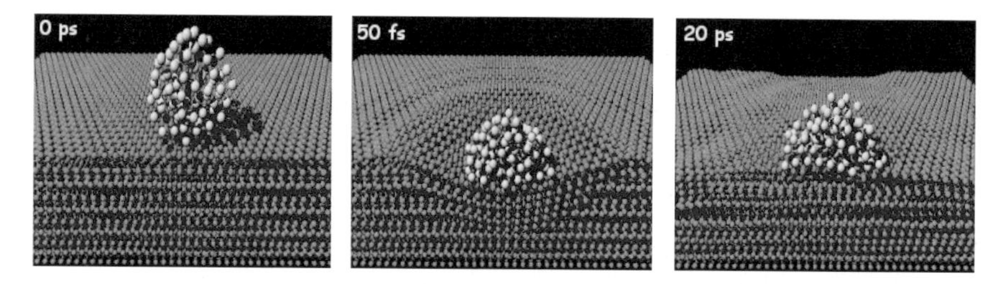

Figure 3.22 Molecular dynamics simulation of the landing of a silver cluster (200 atoms) on Graphite (100) [195]. Surface temperature 77 K. Perpendicular kinetic energy of the cluster 1.3 eV per atom. Reprinted with permission.

dimensional structure of the free clusters is conserved on the surface, although only to a limiting value of energy of incidence. The cluster height on the surface corresponds to the gas phase value up to energies of incidence of 0.800×10^{-19} J (0.5 eV) per atom. Thereafter the relative height decreases: the cluster is deformed into an ellipsoid.

Figure 3.22 shows this effect for the example of a silver cluster of 200 atoms landing on a graphite surface. Apparently the cluster is strongly deformed and becomes ellipsoidal, but it is not destroyed. However, this simulation is very simplistic since it uses less well known, classical interaction potentials (silver-silver, silver-carbon and carbon-carbon). The exact shape of the clusters on the surface depends besides the energies of incidence also on the plastic material constants and the chemical composition of the surface, especially the surface tension or surface free energy. Measurements show that the deformation of the cluster on the surface (i.e. its final contact face) is smaller than expected from the simulations [197].

One possibility to minimize deformation of the cluster upon landing as well as the chemical interaction or aggregation on the surface is to use a rare gas intermediate layer ('soft landing') [198]. This can be, for example, a monolayer xenon, which is adsorbed at low temperatures (50 K). The clusters are deposited on this layer, which is subsequently thermally desorbed. By variation of the xenon layer thickness, size and density of silver clusters on Si(111) can be varied [199]. Selforganized arrays can also be fabricated [200].

Size selection of clusters via molecular beam methods is technologically challenging and usually results in small number densities of clusters. The generation of macroscopic amounts (milligrams) of size selected clusters is a non-trivial task, which has been achieved only for a few different cluster types. Examples include C_{60} clusters or other fullerenes as well as specific gold clusters such as gold-55, which are fabricated chemically in solution via ligand stabilization [201]. Because of their large surface energy, clusters have the tendency to melt with each other once they are close enough on the surface or in solution. In order to obtain a large number density of individual clusters this melting process has to be avoided by coating the clusters with protective layers of inert molecules. These are in many cases organic molecules such as alkane thiols or phenylphosphine, but can also be plain carbon cages [202].

Deposited on a surface, the organic molecules of the protective layer can be used for the formation of ordered cluster rows along step edges or other lithographically defined defect structures on the surface. This has been demonstrated for decane thiole passivated gold clusters on silicon dioxide surfaces [203]. Super lattices of gold clusters (3.7 nm diameter, generated monodisperse from colloidal solution) netted by decane thiole molecules, have also been fabricated.

Finally we note that metallic, semiconducting or dielectric clusters can also act as the core for nano-shelled particles [7]. These nanoshells are usually generated from a solution, including, for example, termination of the cores with amines, attachment of metal colloids and plating the colloids with additional metal. Such nanoshells are especially interesting for optical and sensing applications since the plasmon resonances shift strongly by varying the metal shell thickness [204] (cf. Section 6.1.1).

Problems

Problem 3.1 Calculate in the Rayleigh limit minimum diameter and depth of focus of a HeNe laser beam, focussed through an objective of $NA = 0.5$ (assume $k_1 = 0.6$).

Problem 3.2 The diffraction limit determines the minimum structure size obtainable via simple projection methods on surfaces. How can one beat that limit?

Problem 3.3 Figure 3.7 shows a quantum corral on a very cold single crystal surface. Could one achieve such a configuration also at room temperature and if not, why not?

7) Such core-shell structures can also be generated from non-spherical nano-particles such as nanorods.

Problem 3.4 In general the thickness of a layer on a surface is given in [nm]. How is this value deduced and why is it better to call it an 'effective thickness'?

Problem 3.5 If you have metallic and dielectric materials at hand but like to obtain specific growth modes: which material combination results more likely in layer-by-layer growth and which in cluster growth?

Problem 3.6 What is the minimum requirement for a long chain molecule in order to form a) a Langmuir–Blodgett film and b) a self assembled monolayer (SAM)?

4
Characterization of Nanostructures

Tailoring objects on a size scale of nanometers in a well defined fashion requires the capability to investigate these objects at the nanometer scale. Direct imaging microscopy and scanning microscopies are first choice among the methods to morphologically characterize nanoscaled structures. More indirect but very sensitive structural data are obtained via linear and especially second order nonlinear optical spectroscopy. Additionally these methods allow one to investigate burried interfaces. Diffraction of electron or atom beams results in additional information about the coherence of the investigated nanostructures. This is important, for example, for the characterization of growth of epitaxial layers or for information about the short range order of the atoms within the macroscopic diameter of the impinging beam. The electronic properties of nanostructures are usually investigated via conventional emission methods such as electron-, X-ray or optical spectroscopy.

Detailed, technologically oriented descriptions of different surface investigation methods – although not directed especially towards nanostructures – can be found in [205,206].

4.1
Optical Microscopy

4.1.1
Simple Light Microscopes

Basic light microscopy via lens magnification has been practiced for over four centuries. Figure 4.1 is a schematic of the set up of a light microscope as used since the sixtheenth century following the work of Zacharias Janssen (1590). The object is imaged via an objective and the image is magnified with the help of an ocular lens. The image of the object appears thus bigger on the human retina as it would if no lenses would be used.

Basics of Nanotechnology: 3rd Edition. Horst-Günter Rubahn

ISBN: 978-3-527-40800-9

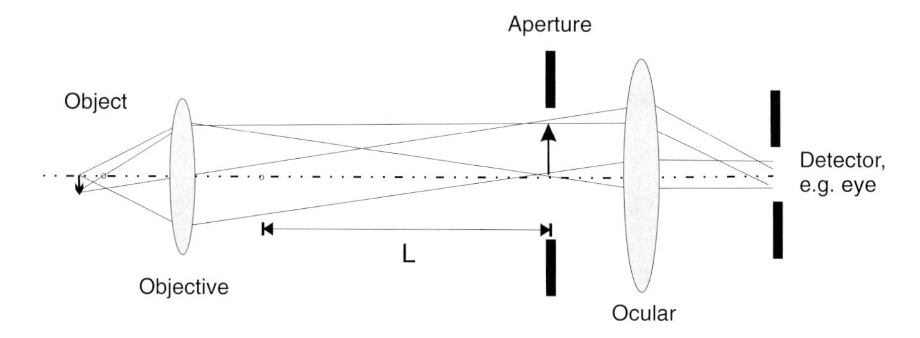

Figure 4.1 Schematic set up of a classical light microscope.

The magnification of this microscope is

$$V = V_{objective} \times V_{ocular}, \tag{4.1}$$

where $V_{objective}$ is the magnification of the objective and V_{ocular} the angular magnification of the ocular. Since a human eye is looking through the microscope ocular, one uses for V_{ocular} the ratio between 'near point'of the human eye and the ocular focus length. The 'near point' is the smallest distance between object and human eye at which the eye is still able to focus. This distance of course depends on human age, which means that one has to define a 'general' near point of 254 mm (10 inch).

It thus follows for the magnification

$$V = -\frac{L}{f_{objective}} \cdot \frac{254\,\text{mm}}{f_{ocular}} \quad , \tag{4.2}$$

where f_{ocular} denotes the focal length of the ocular and $f_{objective}$ the focal length of the objective. L is the 'tubular length', that is, the distance between the focus point of the objective and the first focus of the ocular (typically 160 mm). A 10x objective is thus an objective of focal length 16 mm and a 10x ocular has a focal length of 25.4 mm. The total magnification of this combination is 50.

The smallest resolvable structure or the smallest resolvable distance between two separately imaged points is given by the wave nature of light. For each image a diffraction pattern is generated, which is shown for a point source in Figure 4.2. It can be described by an Airy function.

The diameter of the Airy disc for an imaging lens of focal length f and diameter D is given by

$$d_{Airy} \approx 2.44\frac{\lambda f}{D} \quad . \tag{4.3}$$

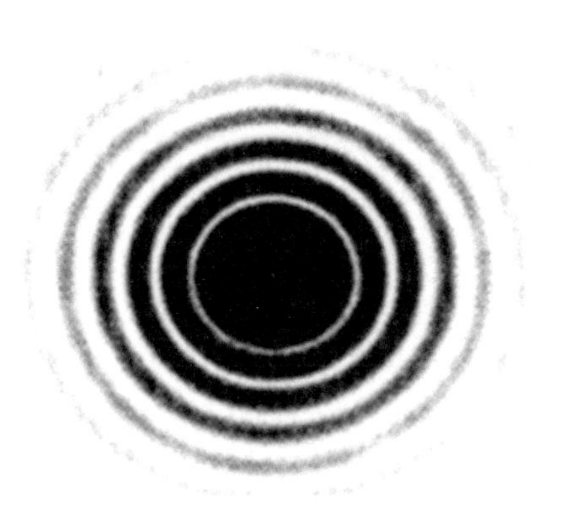

Figure 4.2 Optical resolution limit: diffraction pattern (Airy disc) and transmission function of a single point. High intensity is colored black. The abscissa is presented in units of $\frac{2\pi}{\lambda} a sin\alpha$ with λ the wavelength of the imaging light, a the aperture radius and α half the acceptance angle, defined in Figure 4.4. 'A' and 'B' represent the intensities of the first ($I_1 = 0.0175$ for $I(0) = 1.0$) and second maximum ($I_2 = 0.0042$). The radial distance value of the first maximum is $\frac{2\pi}{\lambda} a sin\alpha = 5.14$, that of the second 8.42.

One can separate two points at distance Δx if the intensity maximum of the first point falls into the intensity minimum of the second point ('Rayleigh–Abbe criterion' [207,208], Figure 4.2). This is the case for

$$\Delta x \approx 1.22 \times f \times \frac{\lambda}{D} \quad , \tag{4.4}$$

assuming that $sin\alpha \approx \alpha$.

The minimum angular distance between two object points $\alpha_{min} = 1.22\lambda / D$ is called the 'resolving power'. The 'resolving power' increases with increasing diameter of the lens and decreasing wavelength of the imaging radiation. In an electron microscope, for example, one uses a wavelength of 10^{-5} of the wavelength of visible light and the resolution is correspondingly higher. If one uses a microscope objective instead of a simple lens, it proves useful to define the resolution limit as

$$d_{\min} = k_1 \frac{\lambda}{NA} \quad . \tag{4.5}$$

The constant k_1 represents a coherence factor, which has values between 0.55 and 0.8, depending on the illumination wavelength. In the case of an Airy function (i.e. a point light source), k_1 =0.61. NA is the 'numerical aperture',

$$NA = n \cdot sin\alpha \tag{4.6}$$

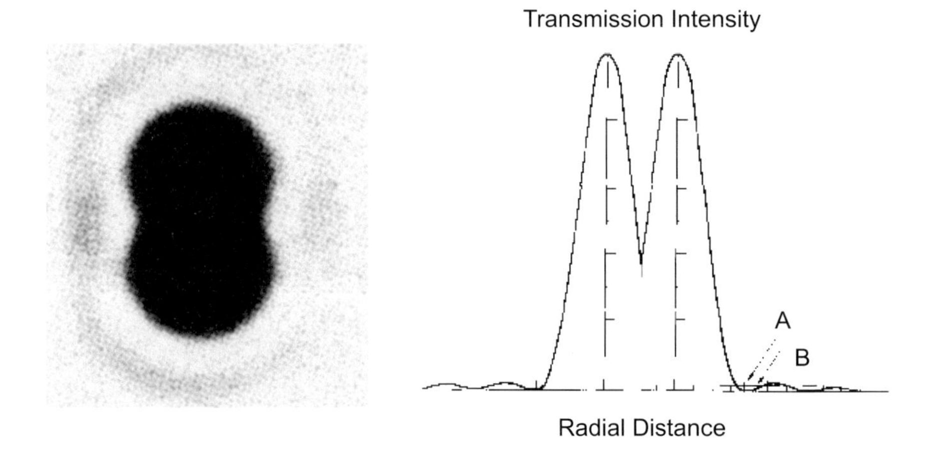

Figure 4.3 Optical resolution limit: diffraction pattern (Airy disc) and transmission function of two neighboring points.

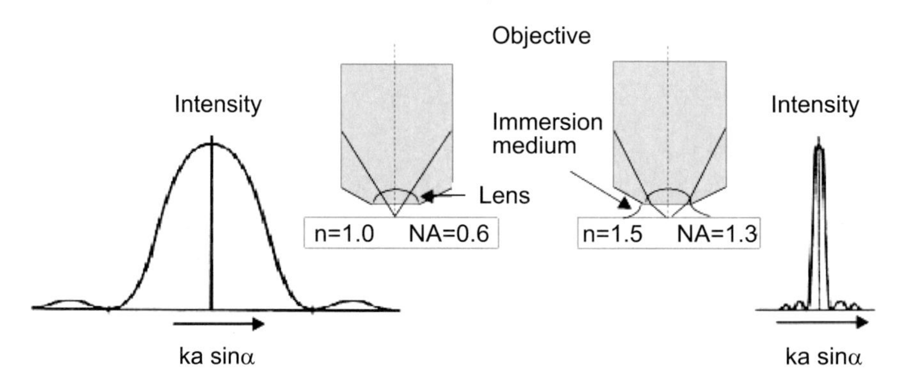

Figure 4.4 Definition of angle α, which is important for the numerical aperture, as well as two examples for the change in the Airy diffraction pattern as a function of NA. The effective viewing angle for the object is 37° on the left-hand-side and 58° on the right-hand-side.

with n the index of refraction of the surroundings and α half of the total angle of the focussed beam (Figure 4.4).

At a magnification of $V' = \frac{0.12 \cdot NA}{\lambda[mm]}$ the limit of useful magnification in a conventional microscope (tubular length 160 mm) is reached, since at that point diffraction and visual limit meet. That means that for a wavelength of 500 nm all magnifications above 240 NA are no longer useful.

The numerical aperture (or more exactly its square) quantifies the ability of an objective to collect light and thus defines resolution, brightness and sensitivity. Values of the numerical aperture above unity are only possible if one takes advantage of total internal reflection. For that purpose a medium with a high index of refraction (an immersion oil, but also other liquids such as water) is placed between the objective and the object slide. This medium enables the light to be diffracted towards the normal of the object lens (Figure 4.4). As a result, a larger number of Airy diffraction orders are transferred into the imaging system, and the contrast increases.

In reality the theoretical resolution limit (Equation 4.5) is in most cases not reached: imaging errors of the optical elements such as astigmatism, chromatical aberration or deformations determine the resolution. An additional limit is the pixel resolution of the CCD camera that is usually applied for obtaining the images. For a typical pixel size of $13 \times 15\,\mu m^2$ one obtains for a 10x objective resolutions of about 1.4 μm – about a factor of ten larger than the nanoscopic dimensions this book is concerned with.

4.1.2
Bright and Dark Field Microscopy

An increase of the numerical aperture of the objective increases the resolving power of the microscope. Illumination with a dark field or evanescent wave condensing optical system results in a similar effect by increasing the optical contrast. Compared to conventional bright field illumination systems (Figure 4.5b) in the dark field system [209] (Figure 4.5a) the central light spot is blocked. Hence the illumination occurs only with light beams which propagate nearly parallel to the surface. The objective thus no longer collects the light beams that have passed the object, but only the light beams that have been reflected from the object. Correspondingly the object appears bright in front of a dark background. The idea for this method of illumination dates back to the year 1903 (H. Siedentopf, R. Zsigmondy).

In the evanescent wave illumination system [210] this principle is optimized by using only surface waves for illumination, that is, only the evanescent part of the electromagnetic field. Since one is able to observe in dark field microscopy, in principle, structures with characteristic sizes smaller than the wavelength of the applied light (visibility limit about $\lambda/100$), this kind of microscopy is also called 'ultramicroscopy'.

As an example for the contrast improvement by illumination in the dark field we plot in Figure 4.6a and 4.6b bright and dark field microscopy images of the same object (a curved nanostructure made from organic material). One can clearly see that the contrast is significantly increased by dark field illumination. The main reason for this drastic im-

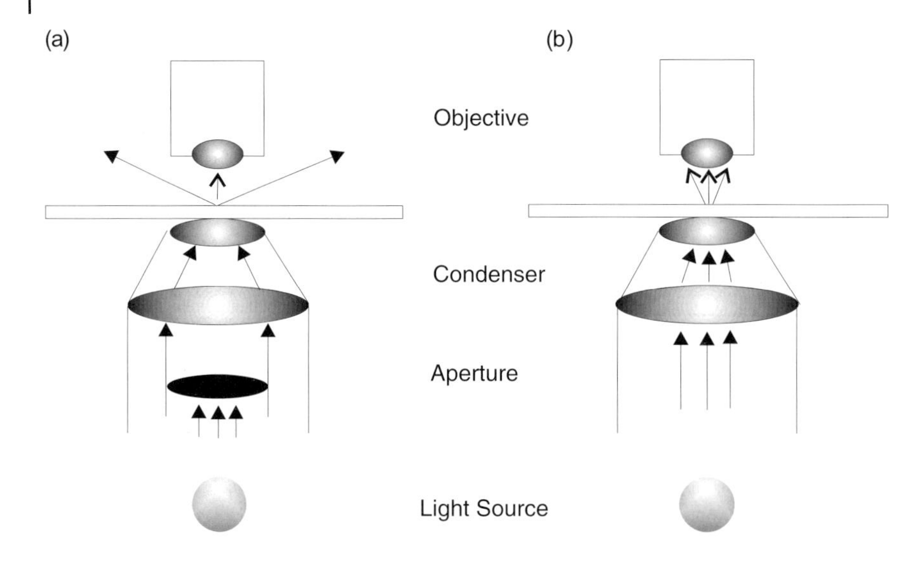

Figure 4.5 (a) Dark field illumination.(b) Bright field illumination.

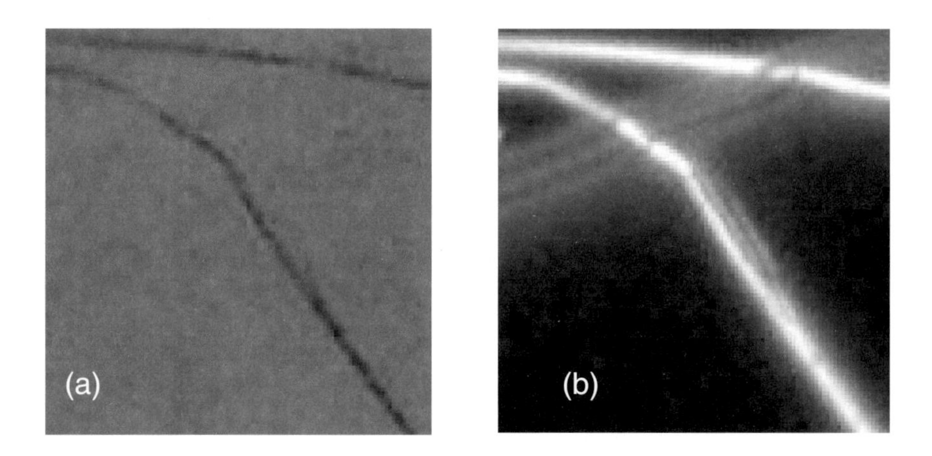

Figure 4.6 (a) Bright field microscopy image of needle-like structures made from organic molecules. (b) Dark field microscopy of the same area. The image size is 20x20 μm^2.

provement is that the structures shown in Figure 4.6 not only possess a small diameter, but are also very flat (thickness about 100 nm). Therefore they absorb only weakly in transmission. From the dark field image a diameter of the wide nanostructure of 1.3 μm is estimated. A scanning force microscopy image (Figure 4.7b) reveals a true diameter of 500 nm.

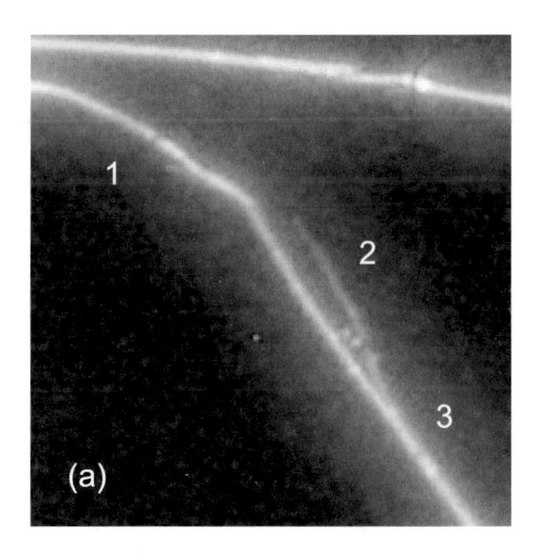

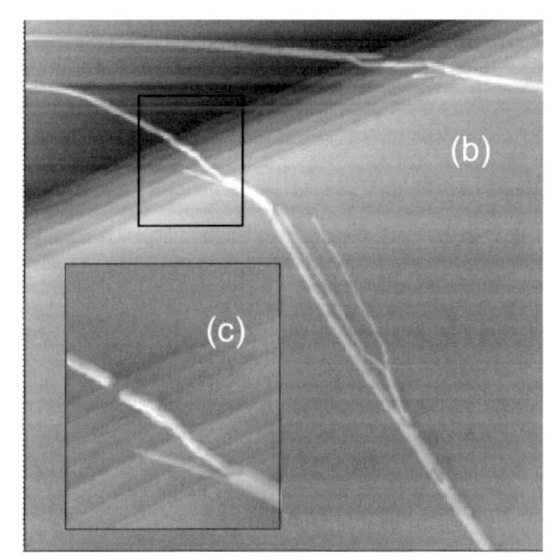

Figure 4.7 (a) Fluorescence microscopy and (b) scanning-force microscopy [211] of the organic needle structures from Figure 4.6. Image size 20x20 μm^2. Typical dimensions are: a break in the needles with width 300 nm ('1'), diameter of a thin needle 300 nm ('2') and a wide needle 500 nm ('3'). (c) Detail image of the break '1' (image size 5x4 μm^2).

4.1.3 Fluorescence- and Phase-Contrast Microscopy

A further enhancement of resolution via contrast improvement becomes possible if one uses fluorescing objects. Figure 4.7a demonstrates this effect, again for the example of the organic nanoneedles from Figure 4.6 which emit intense blue light after excitation in the UV spectral range. The supporting substrate (mica) is not emitting, which results in maximum contrast if one uses an optical filter that allows only the blue light to reach the CCD camera. The apparent diameter of the wide nanoneedle ('3') is about 800 nm. Again, the true diameter as determined from the force microscope image (Figure 4.7b) is 500 nm. In Figure 4.7c a break in the needle has been investigated, which is correlated to its growth over a step edge of the mica support. The width of the break is 300 nm, which is of the same order of magnitude as the small needle structures.

As seen, it is possible to investigate objects with dimensions of a few hundred nanometers using fluorescence microscopy. However, where an exact size measurement is required, even at this size range, other methods must be used, for example, scanning-force microscopy.

In the case of biological objects, such as cytoplasm, another problem is that the absorptivity is very small, similar to that of the surrounding water. The indices of refraction, however, are slightly different

($n_{water} = 1.33$, $n_{cytoplasm} = 1.35$), which results in a delay of the incoming light wave into the cytoplasm. Following transmission through the object to be imaged, a phase difference exists with respect to the surrounding medium.This phase difference can be used for contrast enhancement, as demonstrated in the 1930s by Frits Zernicke [212]. In more detail, a ring aperture is placed between the illumination lens and the object slide, which stops the direct light beam. A phase ring behind the objective reduces the diffraction term of zeroth order in order to increase contrast. This ring results in destructive interference of either the light waves that have been delayed by the object or of those that have not been delayed. In the former case the object appears dark in front of a bright background, in the latter case the background is dark.

The Rayleigh–Abbe limit makes it impossible to separate two objects that have distances of less than $\lambda/2$. However, one is able to determine the *position* of an isolated object with much higher precision. For that purpose one scans the object spatially and records the signal intensity. The precision with which one can determine the position, δx, is given approximately by

$$\delta \propto \frac{\Delta x}{SN\sqrt{n}} \quad , \tag{4.7}$$

where Δx is the width of the spatial intensity curve, SN the signal-to-noise ratio and n the number of measured points. An example for the use of this method is the investigation of the influence of the local environment on the properties of individual quantum dots [213].

4.1.4
Confocal Microscopy

Via scanning imaging – in the optical spectral range especially via near field microscopy – a much higher resolution is generally available compared to conventional microscopy. However, for example in biology or biophysics information concerning the interior of biological objects (e.g. living cells) is just as important as the plain surface information. In order to improve the depth information above what conventional microscopy can offer a variant, namely confocal microscopy, has been developed [214]. Due to significant progress in dye photochemistry, fluorescence markers on proteins, DNA and cell organells also play an ever more important role. Such fluorescing probes allow one to perform confocal microscopy with multiphoton processes, which means that i) UV-transparent optical components are no longer necessary (excitation typ-

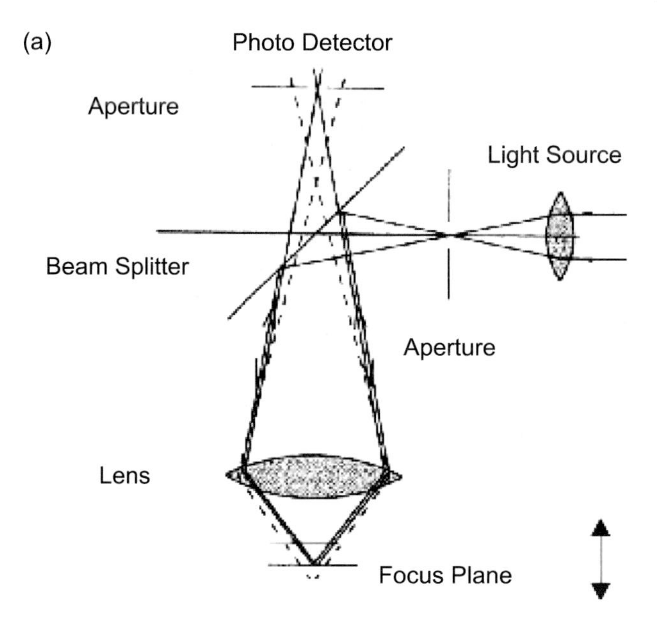

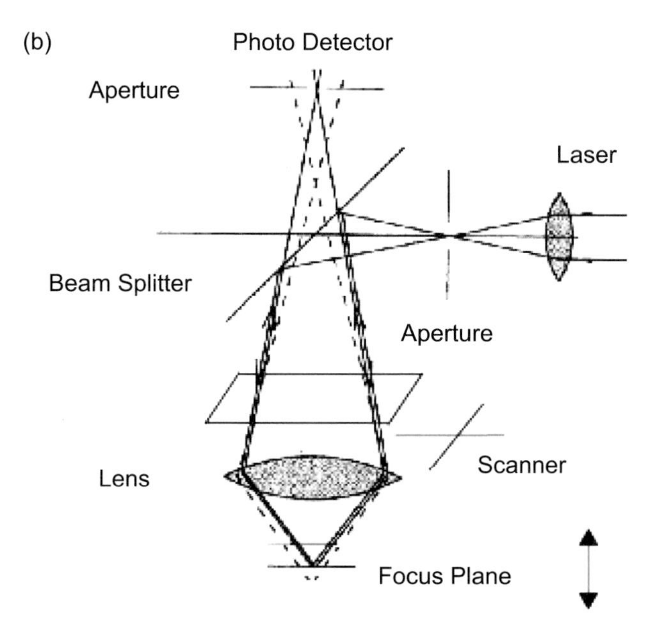

Figure 4.8 (a) Schematic of a confocal microscope. The rays drawn with straight lines are for an object inside the aperture diameter, the dashed lines are for an object outside. (b) Modification of the confocal microscope to a 'laser-scanning microscope' which is often used for two-photon microscopies.

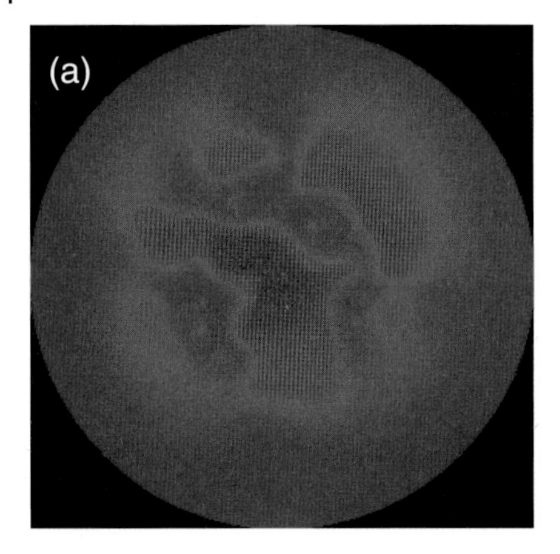

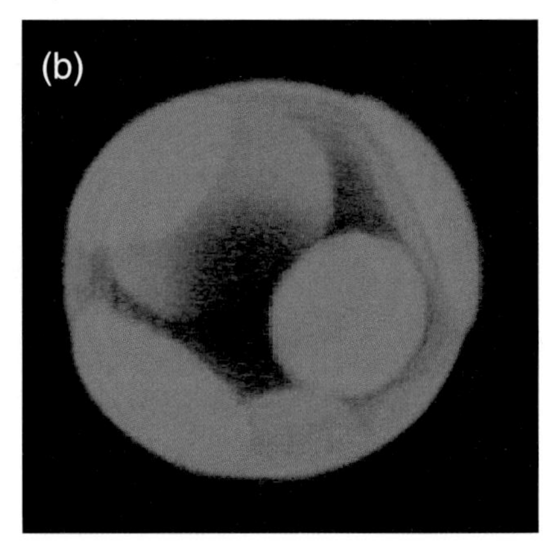

Figure 4.9 Confocal fluorescence microscopy of GUVs (giant unilamellar vesicles) [217]. The diameter of the vesicles is 25 μm and they are marked by fluorescence dyes. In (a) the GUV consists of POPC/DPPC, in (b) cholesterol is mixed in. In (a) liquid (bright red) and jelly-phases (dark) coexist. In (b) ordered (dark) and disordered (bright green) liquid areas coexist. The influence of cholesterol becomes visible at the phase borders [218].

ically in the near infrared) and ii) the resolution further increases under appropriate conditions [215][1].

Figure 4.8a is a schematic of a confocal microscope. The sample is illuminated from above through an aperture. In front of the photodetector another aperture is placed, which makes sure that only those rays are detected which stem from the focal plane of the objective lens. In a confocal microscope the effective focus acts as a three-dimensional probe, which can be scanned through a transparent object. As shown in Figure 4.9 with the help of vesicles made from mixtures of organic materials, three-dimensional images of various structural phases in and on the surface of the vesicles can be obtained. In Figure 4.9 fluorescence dyes have been used for that purpose.

The possible resolution is demonstrated in Figure 4.10a with the help of fluorescing spheres with 110 nm diameter. The distortion of the spheres is due to the fact that the axial resolution is of a factor three to four smaller than the lateral resolution.

In order to increase the axial resolution one might illuminate the sample coherently from the rear. 'Coherent' means that a fixed phase relationship exists between front and rear side illumination. For example, one might illuminate from both sides with light from the same laser. This

1) A non-trivial statement: multiphoton excitation of fluorescence markers decreases the resolution to first order since the excitation wavelength is increased [216].

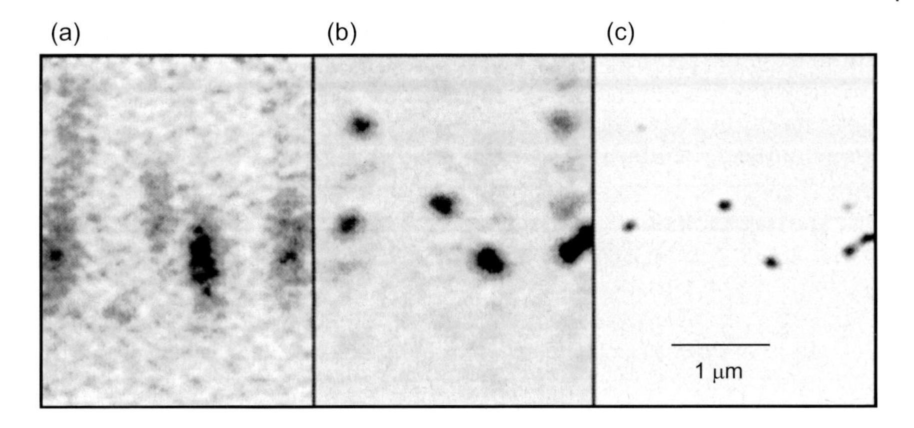

Figure 4.10 Image of a distribution of fluorescing spheres with diameter 110 μm via (a) confocal microscopy, (b) 4π confocal microscopy and (c) 4π confocal microscopy with image restoration. The image is shown in (x, z)-direction, that is, axially with the imaging focus [215]

'4π confocal microscopy' strongly reduces the axial side-maxima of the scattering function and increases thus the axial resolution (Figure 4.10b). In the ideal case (complete illumination with solid angle 4π, which is of course with two lenses of finite diameter not possible) one would completely eliminate the side-maxima. Subsequent image reconstruction allows one to simulate that scenario (Figure 4.10c) [215]. Axial and lateral resolution are then of the order of 100 nm.

Another improvement of the far field resolving power, important for the implementation in nanobiology, employs two short pulse lasers [219]. The idea is to use the first laser pulse for excitation of the probe (e.g. in the yellow spectral range at 558 nm) and the second, picoseconds delayed pulse, for quenching of the generated fluorescence (e.g. in the red spectral range at 766 nm). Quenching occurs in the overlap range of both pulses with the quench (STED – stimulated emission depletion) pulse being phase modulated such that a donut profile[2] around the excitation pulse is generated. As a result only a fluorescence image of the inner core of the excitation spot is generated and one is able to discriminate between objects which are separated less than $\lambda/11$. If one combines this trick with 4π confocal microscopy, lateral and axial resolutions below 40 nm become possible [220].

2) A donut-profile has an intensity minimum in the middle, surrounded by a ring of high intensity. Such a profile can be easily generated by the overlay of two transversal laser modes, one of which possesses no intensity in the beam center.

4.1.5 Brewster Angle Microscopy

If polarized light is refracted at an interface between a substrate and a surrounding medium, p- and s-polarized parts will be reflected differently, depending on the angle of incidence. Linear p-polarization means that the electric field vector oscillates in the plane of incidence and s-polarization that it oscillates perpendicular to that plane. At an ideal interface, where the index of refraction changes from substrate to surrounding, p-polarized light, incident under the 'Brewster angle', will not be reflected. The 'Brewster angle' is the angle where the reflected beam and the beam that is refracted into the substrate are mutually perpendicular.

At a real interface reflection of the p-polarized light will be minimized, but it will not totally vanish. The reasons for this behavior are surface roughness, the finite thickness of the interface (i.e. $n = n(z)$, where z denotes the coordinate normal to the surface) and anisotropies of the monolayers on the interface. This effect is taken advantage of in 'Brewster angle microscopy', BAM [221–223]. A densely packed monolayer of amphiphilic molecules changes the index of refraction in the interface, $n(z)$, along a thickness $l = 2\,\text{nm}$. Both $n(z)$ and l depend on the different phases of the monolayer (e.g. solid vs. liquid), which means that contrast is generated via the different depths of the Brewster minimum. If one scans, for example, a polarized green light source (e.g. an argon ion laser) under the Brewster angle over the surface, the different monolayer phases will generate a bright-dark image.

BAM is thus a two-dimensional variant of ellipsometry, which allows one to determine layer thickness and optical properties of adsorbed monolayers. If those are known, detailed morphologic properties of adsorbate films (molecular orientations or phase transitions) will be revealed.

4.2 Scanning Microscopies

The resolution in direct imaging methods can be improved by decreasing the wavelength of the imaging beam. This leads to imaging methods based on, for example, electrons instead of photons.

4.2.1 Electron Microscopy

The transmission electron microscope [224] is constructed analogous to the light microscope (Figure 4.1). The light source, however, is replaced by an electron source, the optical lenses by magnetic electron lenses and a photophorescence screen or an electron multiplier serves as detector.

For an acceleration voltage of 100 kV the imaging electrons possess a wavelength of 0.0038 nm. This wavelength is a factor 10^5 smaller compared to the wavelength of visible light, and the resolution should be correspondingly higher (Equation 4.4). Indeed, as early as 1956 atomic lattice planes with a distance of 1.2 nm within a few ten nanometer thick single crystalline layers could be resolved. Nowadays lattice planes with separations of below a tenth of a nanometer can be routinely determined.

One problem of this method is that the necessary electron energy for transmission increases strongly with increasing thickness of the material owing to scattering losses – it is, however, difficult to work with energies larger than a million electron Volts. Thus, transmission electron microscopy is applicable only to very thin substrates, the properties of which might be different from the properties of the corresponding bulk matter.

A second approach is scanning electron microscopy. In this case an electron beam, which has been focussed via magnetic coils to roughly 10 nm diameter, is scanned via deflection coils along the object. The electrons, which are either backscattered from the object or generated via secondary processes, induce a voltage in an electron multiplier, which in turn regulates the brightness of a synchronously scanned electron beam on a fluorescence screen[3]. As a result, a magnified image is projected onto the fluorescence screen as the object is scanned.

In the scanning electron microscope the possible resolution for thick objects is limited to about 1 to 2 nm due to, for example, electron diffusion which widens the focussed electron beam. With decreasing object thickness the resolution increases and the limits are then set by the charging of the objects, imaging errors of the lenses[4] as well as any failings in the mechanical stability of the instrument.

3) In modern versions of the SEM, the fluorescence screen is of course replaced by a computer screen.

4) The most important imaging error is the spherical aberration, that is, the focusing of non-paraxial rays in front of the paraxial rays. This error cannot be in general eliminated if one applies rotationally symmetric lenses [225].

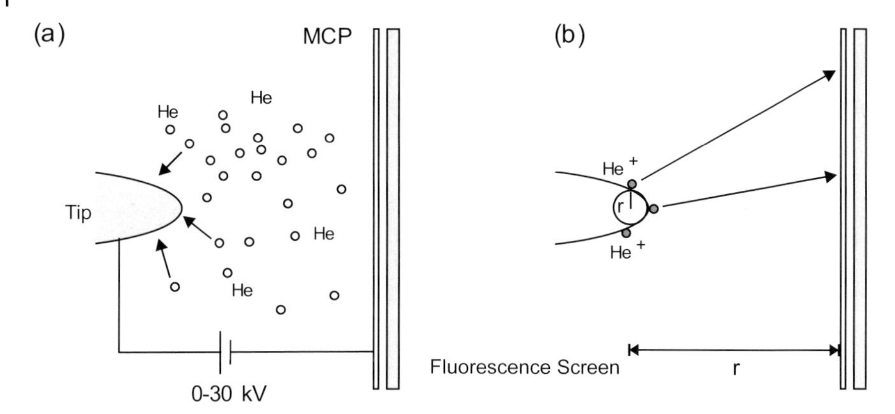

Figure 4.11 Field ion microscope (FIM). (a) Attractive force between tip and helium atoms due to the polarization of the atoms by the electric field. (b) Repulsion and acceleration of the positively charged helium ions onto the fluorescence screen behind a multi-channel plate (MCP) amplifier.

Very high resolution, down to the sub-Ångstrom regime, is obtained via non-scanning electron holography[5] [226, 227] or low temperature field ion microscopy ('FIM', Figure 4.11)[228]. In FIM the sample is formed as a tip. Helium atoms from a diluted buffer gas (partial pressure 0.013 Pa (0.1 mTorr)) are accelerated towards the tip by a high potential difference of a few thousand Volts. Half a nanometer in front of the tip, field strengths of a few $10^{11}\,\mathrm{Vm}^{-1}$ ionize the atoms. Subsequently they are accelerated towards a fluorescing screen, where they induce fluorescence. Usually a multichannel plate multiplier is mounted in front of the fluorescence screen in order to increase the electron current. Due to the large acceleration voltage the helium ions propagate on nearly straight trajectories. Thus a tip with a radius r of about 10 nm is amplified onto the screen at a distance $\Delta r = 10\,\mathrm{cm}$ by a factor $10\,\mathrm{cm}/10\,\mathrm{nm} = 10^7$. This magnification is sufficient for subnanometer resolution and for discriminating individual atoms. Diffusional motion of individual atoms has been investigated this way, although the size of the diffusion area was obviously very limited.

5) Analogous to optical holography, amplitudes and phases are measured in electron holography after interference of the image wave with a plane reference wave. The image wave is thereafter numerically reconstructed, which allows one to correct for imaging errors induced by the electron lenses.

(a)

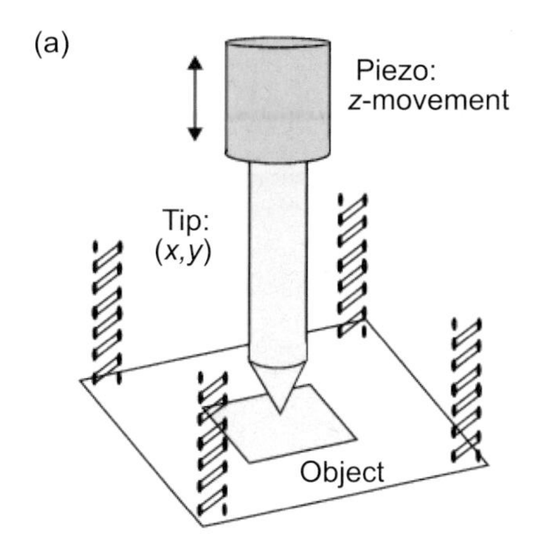

(b)

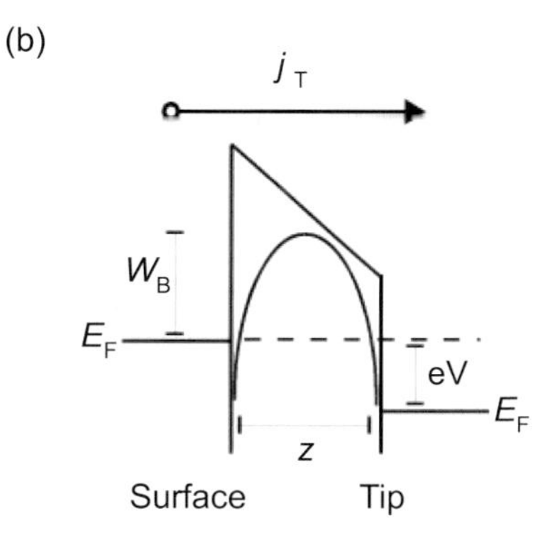

Figure 4.12 (a) Set up of a STM. A conducting tip is mounted on a (x, y, z)-adjustable Piezo-stack. The distance to the surface, z,is regulated by applying a voltage V between tip and surface and measuring the tunnel current j_T. Another voltage is used to scan the tip in (x, y)-direction along the surface. (b) Energy level scheme for a tip with negative bias, with Fermi-energies E_F, effective work function of the barrier, W_B and distance between tip and surface, z.

4.2.2
Tunneling and Force Microscopy

With the help of scanning tunneling microscopes (STM) (Figure 4.12) a vertical resolution in the sub-nanometer or even sub-Ångstrom regime can be achieved on surfaces over large spatial areas [229, 230]. The STM [231] uses changes in the tunnel current j_T between a conducting tip and a conducting surface to monitor the distance between tip and surface. The tip is mounted on piezoelectric actuators, which allow a reproducible movement in the subnanometer regime.

Piezoelectric ceramics change their size upon application of a high voltage. Their advantage with respect to other actuators is that they allow a movement without hysteresis since their extension builds on a continuous atomic movement. In addition they are very stiff (about 20% of the stiffness of stainless steel), and their response time is limited only by the electronics of the controller. The precision of translation is given by the noise of the movement amplifier and is of the order of $0.4\ \text{pm}/\sqrt{Hz}$ [232]. A significant disadvantage, especially for the use in high temperature STMs, is the low Curie temperature above which the field-induced polarization is lost.

The tunneling current for given voltage V between tip and surface is

$$j_T \propto \rho(r, e_F) V \exp\left(-2\frac{\sqrt{2m(W_B - E)}z}{\hbar}\right) \qquad (4.8)$$

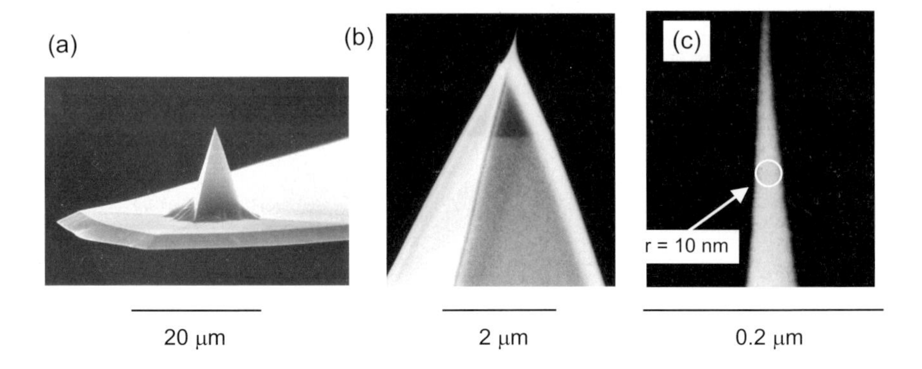

Figure 4.13 Scanning electron microscopy images of silicon tips for force microscopy (AFM). (a) Silicon support with integrated tip. The typical radius r is $10\,\mathrm{nm}$. (b) 'Super sharp' tip, and (c) close up of the super sharp silicon tip. Minimum radius about $2\,\mu\mathrm{m}$. Reprinted with permission from [233]. Copyright 1999 Nanosensors.

and depends exponentially on the distance to the surface z ('barrier width'), the square root of the electron mass m and the difference between height of barrier (W_B) and electron energy, E. The pre-factor $\rho(r, e_\mathrm{F})$ describes the spatial electron density distribution near the Fermi level; this is in fact the quantity that is imaged. Depending on whether the tip is biased positively or negatively with respect to the surface one observes either the highest occupied molecular orbitals (HOMO) or the lowest unoccupied molecular orbitals (LUMO). The exponential distance dependence results in a strong change of tunneling current even for a small change of distance, which results in a high topographic sensitivity of the method.

An important prerequisite for a successful use of scanning microscopies is, in addition to fast and reliable electronics, an efficient vibration damping (in Figure 4.12 symbolized by springs). Modern STMs are simple to use under ambient air conditions and do not require advanced knowledge or specific working conditions [234]. More problematic is their use under extreme conditions, such as very high [235] or very low temperatures [236], very high pressures [237] or in an extremely good vacuum. However, even for these conditions many commercial solutions are currently available.

The measurement of a tunneling current requires that free charge carriers exist within the investigated substrate. While extensive STM studies on metal and semiconductor surfaces are available (e.g. [238]), insulator surfaces cannot be directly investigated with a STM. An exemption are

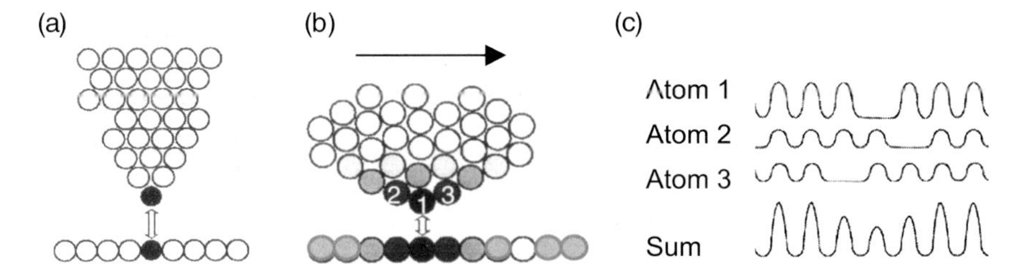

Figure 4.14 A scanning tunneling microscope senses local atom interactions (a), while the cantilever deflection of a force microscope shows a less strong distance dependence (b). A defect on the surface (e.g. a missing atom) is therefore hidden easily in the periodic background (c): the local resolution of the AFM is much smaller compared to that of a STM.

very thin insulating films such as organic films on metal surfaces (Figure 3.13), which allow electrons to tunnel through them. The STM then images the distortion of the electron density distribution of the plain surface caused by the presence of the organic overlayer.

In order to achieve routinely at least nanometric resolution on insulators scanning force microscopes (AFM, 'atomic force microscope') can be applied. The AFM [239] takes advantage of the van der Waals force (or other attractive or repulsive forces close to surfaces), which leads to vertical and torsional distortions of the tip, which is mounted on a cantilever (Figure 4.13); this force has a cubic distance dependence to the surface [240].

For a quantitative analysis of AFM images in the nanometer range one has to take into account that the AFM tip also has a diameter of a few to a few tens of nanometers (Figure 4.13). Since the interaction force that leads to a deflection of the cantilever is exerted by more than one atom at the tip, the achievable resolution is much less compared to a STM (Figure 4.14). It is possible to image periodic structures with atomic resolution, but not isolated details. The problem becomes obvious when one considers the reflection of the shape of the tip in structural details of an high-resolution AFM image (Figure 4.15). In order to correct this effect, the measured image has to be deconvoluted with the effective tip shape [241]. Modern image processing programs allow for a 'blind deconvolution', however, a more serious approach would be to determine the shape of the tip by scanning a sample surface with very well known topology (e.g. a periodic array of nanometric particles on a surface) and subsequent deconvolution. Even under optimum conditions this method is non-trivial and does not necessarily provide unique results. Simulta-

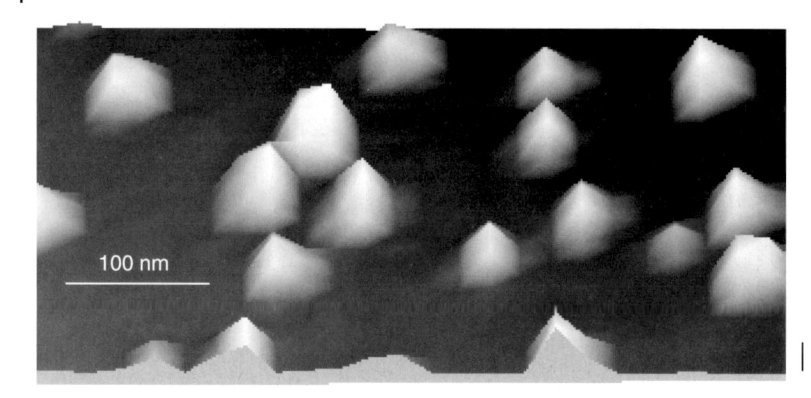

Figure 4.15 High resolution AFM image of aggregates made from light-emitting organic molecules, adsorbed onto an insulator surface, which could act as quantum dots. The shape of the imaging tip of the AFM is reflected in the shape of the aggregates, the morphology of which cannot be exactly determined.

neous measurements using an alternative technique (e.g. fluorescence microscopy for larger structures) are then useful.

Force microscopy is currently for biological basic research a very important experimental tool [242].

Besides STM and AFM a multitude of modifications of these high resolution scanning probe microscopies have been developed since the 1980s. Among those are electrostatic force microscopy, scanning thermal microscopy, magnetic scanning microscopy, friction microscopy, spin-polarized STM etc. Overviews are given in [243–245].

4.2.3 Near Field Microscopy

A successful synthesis of AFM microscopy and conventional optical microscopy is near field microscopy (Figure 4.16). The diffraction limit of optical microscopy can be beaten by placing the imaging aperture or the detector itself into the *near field* of the reflecting or light emitting object. The idea, to apply an aperture with a diameter which is smaller than the wavelength of the imaging light, dates back to Synge [246]. An experimental realization followed in the 1970s in a reflection-SNOM ('scanning near field microscope') for centimeter waves [247] with a resolution of $\lambda/15$. Ten years later followed a variant in the visible spectral range (λ = 488 nm), the SNOM ('scanning near field optical microscope') with a resolution of better than 25 nm or $\lambda/20$ [248–250]. At the end of the 1980s

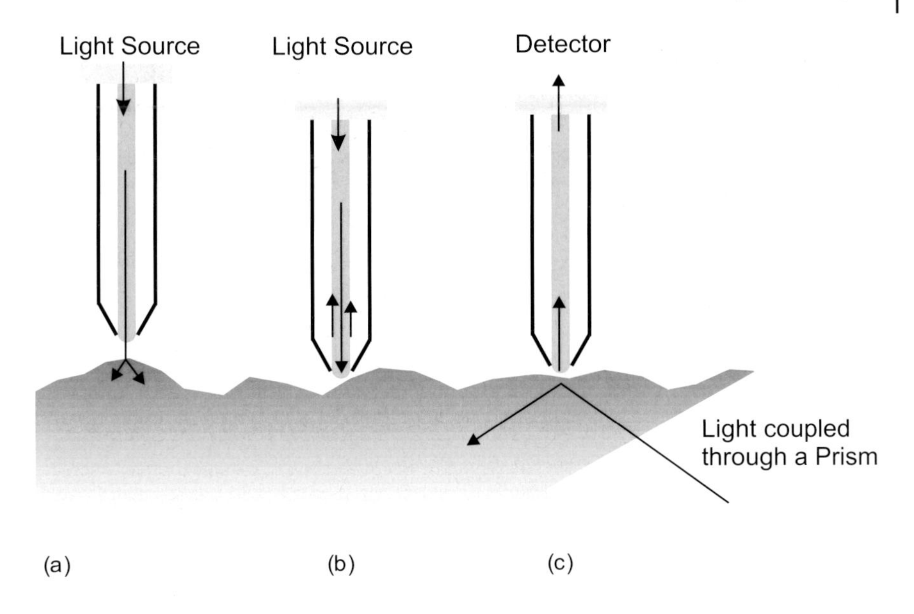

Figure 4.16 Imaging methods using near field microscopy: (a) SNOM in transmission, (b) SNOM in reflection, (c) PSTM.

a new modification of the SNOM was constructed, the PSTM ('photon scanning tunneling microscope', Figure 4.20) [251].

An important prerequisite for a practical realization of all these ideas is to be able to scan the imaging object (e.g. the aperture) with a few ten nanometer diameter reproducible and with sufficient resolution at a distance of a few nanometers along the surface. The development of scanning microscopies as described in Section 4.2.2 has provided this technology and thus made the development of the SNOM possible [252]. Instead of using an imaging aperture, in most cases, the tip of an optical fiber is used, which is coated with aluminum except for a small hole at the tip. Consequently, the collected photons can be directed to a photomultiplier, which increases the sensitivity. The optical fiber is usually mounted on a piezoelectric stack, and similarly to an AFM the shear forces are measured during scanning for example, via changes in the resonance oscillation of the optical fiber holder. One obtains, with nanometer resolution, simultaneously topographic and optical information concerning the surface. The resolution limit is given by the penetration of the light into the aluminum coating (the 'skin depth'), which limits the effective diameter of the imaging aperture and, thus, the resolution to about 10 to 30 nm (Figure 4.17).

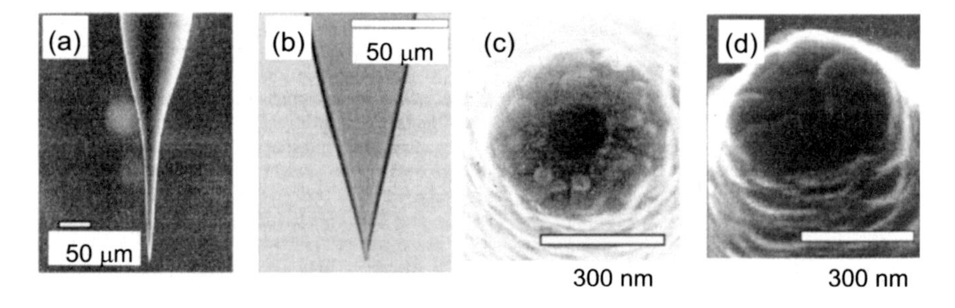

Figure 4.17 Side and front view of aluminum coated optical fibers. The images (a) and (c) as well as (b) and (d) are related. (b) is a light microscope image, (a), (c) and (d) are SEM images. The SNOM aperture is visible as dark circle in (c). Reprinted with permission from [253]. Copyright 2000 American Chemical Society.

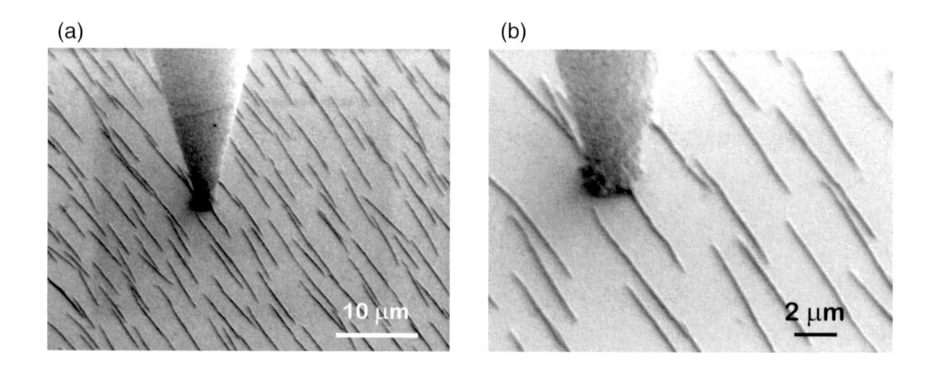

Figure 4.18 (a) SEM image of a SNOM tip above an array of nanofibers. Reprinted with permission [254]. (b) Higher resolution image of the same tip.

Figure 4.18 is an SEM image of a metal-coated SNOM tip on top of an ensemble of light-emitting nanostructures. The size difference between detector and nanoscopic objects is apparent – this difference leads to a low imaging sensitivity and a large uncertainty in localization. The light-emitting objects are situated on an electrically insulating surface in order to minimize the disturbance of the optical properties. In order to avoid charging effects the SEM image had to be taken in a low water pressure atmosphere. This degrades slightly the resolution.

In Figure 4.19 topographical and optical images of such nanoscopic objects ('nanofibers' or nano-waveguides) are shown. The topography (4.19a) is obtained simultaneously with the optical signal since the SNOM tip is scanned at constant distance of a few nanometers along

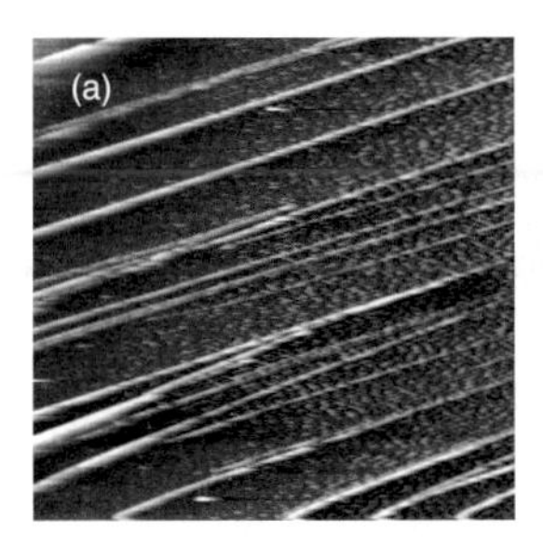

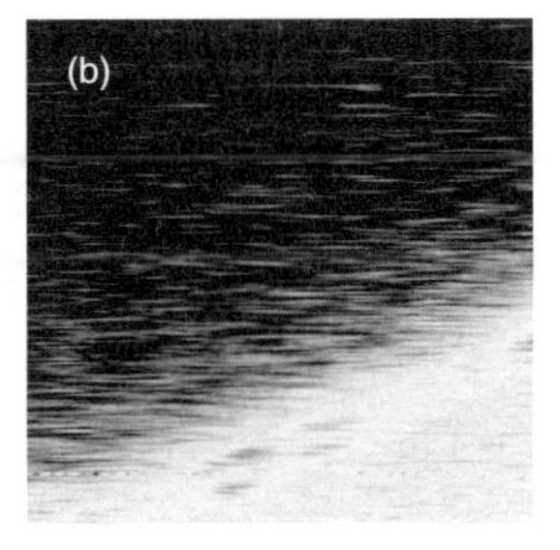

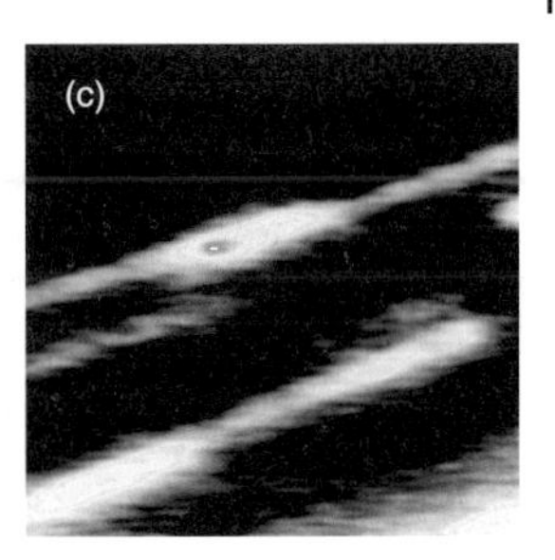

Figure 4.19 SNOM images ($40 \times 40\ \mu m^2$) of nanofibers. Images (a) and (c) have been obtained simultaneously. (a) shows the topography of the aggregates, (c) the optical response following excitation of the nanofibers. In (b) the nanofibers have been excited perpendicularly to their long axes: obviously they do *not* guide light in this direction [255].

the surface. In order to obtain an optical signal the nanoobjects have been excited with UV light. Depending on whether the excitation is perpendicular to the long needle axis (4.19b) or parallel (4.19c), one detects luminescence from the nanofibers which are inside the excitation light (4.19b) or one detects guiding of the light spot along some of the nanowaveguides (4.19c). Wave guiding is seen here as luminescence of the needles along their whole axis since the SNOM collects the signal in the near field of the nanofiber – the SNOM acts as an additional defect of the needles, which – similar to the breaks shown in Figure 3.17 – results in light scattering.

It should be possible, in principle, to improve the resolution of the SNOM to molecular resolution by avoiding the damping via the metal coating. This collection mode is then called 'apertureless' near field optics [256]. A possible approach is given by the PSTM (Figure 4.20) using excitation of the sample surface via an evanescent wave and detection via 'dipping' the fiber tip into the near field of the surface (Figure 4.19 has been obtained that way). As an alternative one might also induce the near field limited field via far field illumination of a strongly scattering object, which is in direct contact with the fiber, which guides the light to the detector. Fiber and scatterer are then scanned at small distances above the surface, which is illuminated in the far field.

While the resolution is no longer limited by the damping aperture, an obvious disadvantage of the method is that small signal intensities have to be measured on top of a high background intensity, given by the illuminated surface. Thus the method works only satisfactorily if the scattering rate of the object that has been mounted at the tip has been

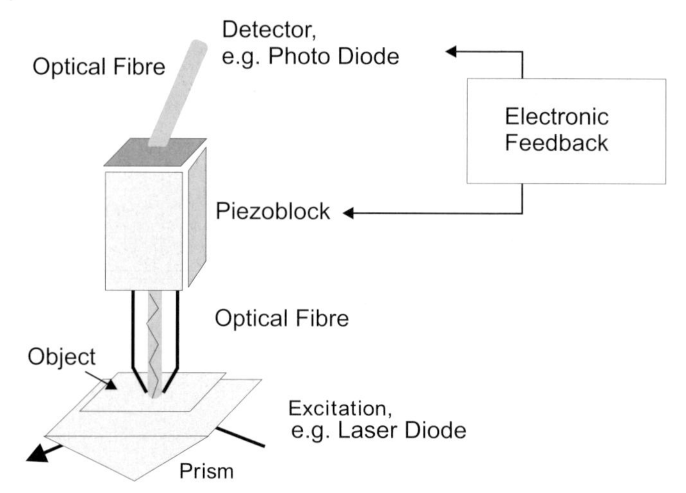

Figure 4.20 Schematic of a PSTM ('photon scanning tunneling microscope'). Here localized electromagnetic fields are detected in the near field of the sample surface via a non-coated dielectric tip.

enhanced, for example, via plasmon excitation (see Figure 6.14 and the corresponding discussion) [257]. An alternative method of contrast enhancement is to use nonlinear optical methods such as optical second harmonic generation, SHG, see also Section 4.3. This has been demonstrated, for example, by SHG from InAlGaAs semiconductor quantum dots on GaAs(001) both in the far field (SH-SFOM, 'SH-scanning far field optical microscopy', [258]) and in the near field (SH-SNOM, [259]) [6].

As a direct comparison with the near field images (Figure 4.19) in Figure 4.21 SH-SFOM-images of nanofibers are presented. The spatial resolution is less good ($\approx 700\,\mu m$) compared with the SNOM, but the contrast ratio signal to background is much better. This allows one to obtain images using different polarization combinations of excitation and emission light. The images a) and b) show the same area but with differently polarized illumination of the fibers. Obviously the optical response of the fibers depends locally on the polarization of the excitation light (see also Figure 4.36 and the corresponding discussion). From the ratio of the light intensities in the polarization combinations I_{sp} and I_{pp} one obtains via

$$\frac{I_{sp}}{I_{pp}} = tan^4\theta \tag{4.9}$$

6) The SH-SNOM works without metal coating of the optical fiber, but the excitation of the sample occurs in the near field via the optical fiber. The corresponding model variant is shown in Figure 4.16a.

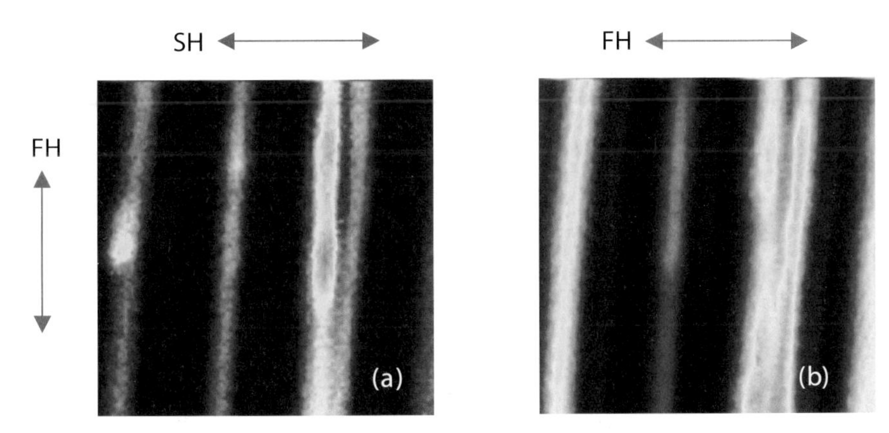

Figure 4.21 SH-microscopy ($10 \times 10\,\mu m^2$) of nanofibers for two polarization combinations. The excitation occurred via a femtosecond laser in the red spectral range, the detection in the blue. (a) Excitation light ('FH') polarized along the needle axis (' I_{sp}'), (b) excitation light polarized perpendicularly to the needle axis ('I_{pp}'). Detection ('SH') polarized perpendicularly to the needle axis [260].

the angle $\pm\theta$ between the axis of the light emitting molecules and the polarization vector of the light. Since the polarization ratio can be deduced directly two-dimensionally from the microscopy images, the orientation of the light emitting molecules along the nanofibers can be determined in an optical manner.

In order to further increase the resolution, the scattering object at the tip could be a single light emitting molecule. In reality single molecules cannot be mounted at the tip of the optical fiber, rather molecular doped host crystals. If one is able to identify spectroscopically individual molecules in the host crystal, they can subsequently be used as single molecule light sources for optical near field microscopy. In Figure 4.22 a set up for an experimental realization of this idea is sketched [261].

As host crystal a para-terphenyl crystal with a diameter of a few micrometers has been used, which has been doped at small concentration (10^{-7}) with terrylene molecules. The tip of the optical fiber has to be kept at very low temperatures (1.4 K) in order to ensure that the spectral fluorescence line width, measured with a laser, is due to the natural line width of individual molecules (a few ten MHz) and not due to collective effects such as coupling to lattice vibrations of the host crystal (Figure 4.23). The spectral response of the individual molecules can then be used for a local analysis of the surface, albeit with relatively low resolution (180 nm) (Figure 4.24). Further tricks (application of an electric field

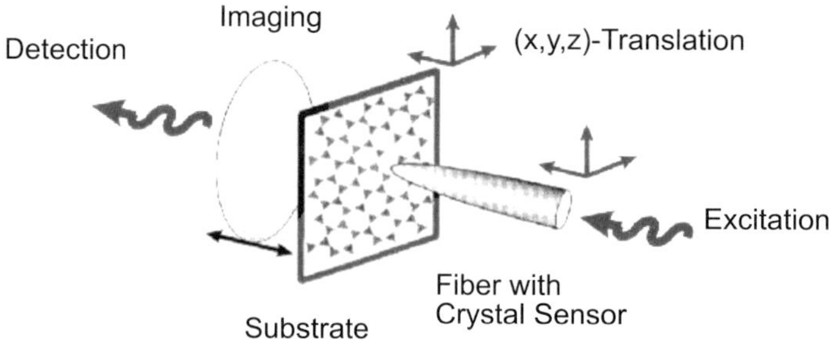

Figure 4.22 Set up for optical microscopy with individual molecules. Reprinted with permission from [261]. Copyright 2000 Nature.

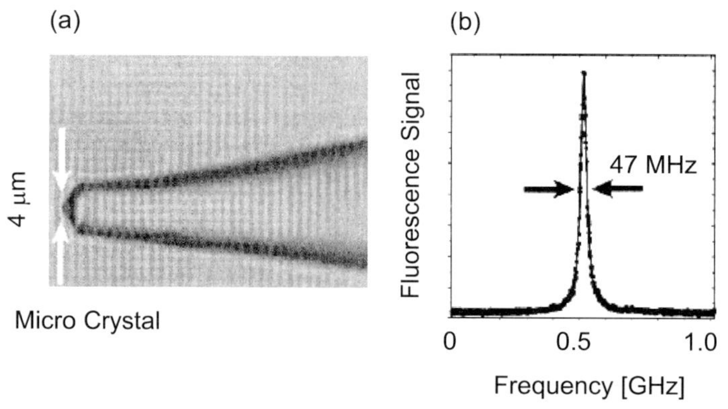

Figure 4.23 (a)Light microscope image of an optical fiber with a p-terphenylene host crystal at the tip, which has been doped with terrylene molecules. (b) Excitation spectrum of the terrylene molecules at $T = 1.4\,K$. Reprinted with permission from [261]. Copyright 2000 Nature.

and use of the Stark shift) allow one to determine at least the position of the investigated molecule to within a few ten nanometers.

For the images in Figure 4.24 the sample consisted of a microscope slide, coated with a hexagonal lattice of 25 nm high triangular aluminum islands. The lattice period was 1.7 μm. In the force microscopy image (Figure 4.24a) two differently oriented aluminum triangles are visible ('i' and 'ii'). The aluminum triangles suppress the optical signal and thus appear in the optical image as dark spots (Figure 4.24b). The differently oriented triangles can be easily separated.

The single molecule light source in fact is a successor of the concept of performing field ion microscopy with molecules that have been marked by laser light in order to localize bonds on a surface [262, 263]. If one deposits a donor dye molecule on the surface and acceptor molecules on the tip of a SNOM or AFM, one obtains not only optical informa-

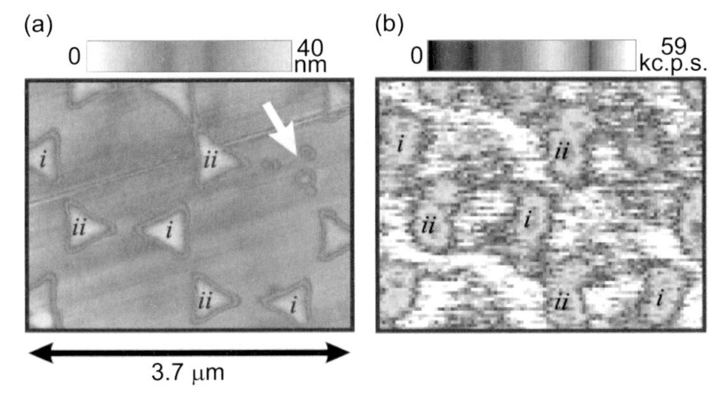

Figure 4.24 Topographic (a) and optical (b) images of a sample coated with terrylene molecules. The optical image has been obtained in the light of a single terrylene molecule. Reprinted with permission from [261]. Copyright 2000 Nature.

tion about the topography but also about resonant energy transfer [264] (FRET, 'fluorescence resonant energy transfer').

4.2.4 New Microscopies

Scanning tunneling microscopy as well as the development of new laser techniques have had a strong influence on the investigation and use of the nano world within the last decade. It is thus sensible to also use laser-induced surface effects and high resolution tunneling microscopy to obtain information about surface properties that is otherwise not available. One example is the application of optical frequency doubling for interface sensitive microscopy [258]. Another approach uses laser light that is coupled into the tunnel connection of a STM and thus induces nonlinear tunnel current components [265]. These components in turn can be used as feedback for the STM ('laser driven STM') and allow one to discriminate between conducting and non-conducting areas on micro structured substrates. Alternatively, one can also determine two-dimensional intensity distributions of elementary optical excitations (e.g. non-localized surface plasmons).

The capture of particles via laser light can further be used for imaging purposes. Eventually it should be possible to capture individual atoms and to image their movements. In order to avoid too strong coupling with the atoms to be imaged, individual photons should be used. This has recently been achieved by use of cavity quantum electrodynamic effects ('atom cavity microscope', [266]). For this purpose cesium atoms have been captured with a laser in an optical resonator. The transmission of a probe laser through the resonator depends on the presence of captured cesium atoms and their position. If one measures this transmission function temporally resolved the position of the atoms can be determined as a function of time with a resolution of about 2 μm.

While this method results in exciting basic research, more broad applications are difficult to foresee. In the following we describe two microscopic methods which are much more application oriented.

X-ray Microscopy

In order to improve the resolution of the microscope, in spite of the diffraction limit, and simultaneously image living objects, it is advantageous to use photons but to decrease the corresponding wavelength – in other words: soft X-rays with wavelengths of a few nanometers should be used. This application of X-rays was discovered shortly after their first observation in 1895 [267].

The most promising wavelength range is between $\lambda = 2.4\,\text{nm}$ (oxygen absorption edge) and $\lambda = 4.4\,\text{nm}$ (carbon absorption edge), since the linear absorption coefficient of water is very low in this range ('water window'), while proteins strongly absorb. In addition the index of refraction of many substances is close to 1 in this range, which reduces the diffuse scattering and results in clear images even for thick samples.

An index of refraction close to 1, however, also means that conventional lens optics no longer work in the X-ray range. Reflection optics for grazing incidence[7] can only be used with strong restrictions due to missing accuracy [268], so that essentially only diffracting elements ('zone plates') are useful [269,270].

A zone plate is a symmetric transmission grating, the grating constant of which decreases radially outwards. Essentially this pattern results from a superposition of a spherical with a plane wave, that is, the pattern is the hologram of a point source. Figure 3.5 shows a conventional zone plate or micro Fresnel lens which can be used for optical imaging. The zone plate consists of concentric rings with outward decreasing width and radius (for first order diffraction)

$$R = \sqrt{r_0 \lambda} \quad , \tag{4.10}$$

where r_0 describes the focal length of the plate and λ the irradiating wavelength. The focus diameter, or the resolution, of the zone plate is dictated by diffraction at the outermost zone as $\delta = 1.22 \cdot \Delta R_n$, where the

7) In the X-ray astronomy the 'Wolters-telescope' bases on such optics.

width is

$$\Delta R_n = \frac{R_n}{2n} \quad . \tag{4.11}$$

Minimum widths of 50 nm have been achieved via holographic illumination techniques and reactive ion etching (cf. Figure 4.26), minimum widths of 20 nm via electron beam lithography followed by anisotropic etching technology.

At present the resolution of X-ray microscopy for biological objects is about ten times higher than with light microscopy. The introduction of zone plates with higher transmission (at present $T = 0.1$ for amplitude zone plates) and new variants such as phase contrast microscopy ($T = 0.2$ for phase zone plates) could significantly increase resolution and application potential in the near future. Additional improvement will be possible through the development of intense sources of soft X-rays from, for example, laser plasmas that could be used instead of synchrotron radiation from storage rings, which requires a lot of experimental effort [271].

Helium Microscopy

Monochromatic, low energy atom beams that are used instead of photons or electrons in imaging microscopy could, in principle, result in improved contrast and larger spatial resolution. In addition low energy atom beams are largely non-intrusive, that is, do not destroy the imaged objects.

Helium nozzle beams are especially useful for the investigation of the structure and dynamics of solid surfaces with minimum influence on the surface since they exhibit large monochromaticity. Usually the helium beams are diffracted at the surface. From an analysis of the diffraction pattern information about the structure of the surface (lattice constant, defects etc.) can be deduced. The inelastic part of the scattering reveals information about the dynamic properties of the solid (phonon excitation or de-excitation). For nanostructured surfaces the interpretation of the diffraction pattern is difficult since long range order is missing – thus the resulting data are not unique in many cases. A 'direct' imaging method is here advantageous.

A scattered atom beam is in fact also *imaging* the surface, but without magnification. In order to obtain magnification, optics is necessary that is transparent for matter beams and that reflects or diffracts the beams[8].

8) Conventional focussing via electromagnetic fields is not possible for ^{4}He-atoms, which exhibit very low polarizability and have no spin.

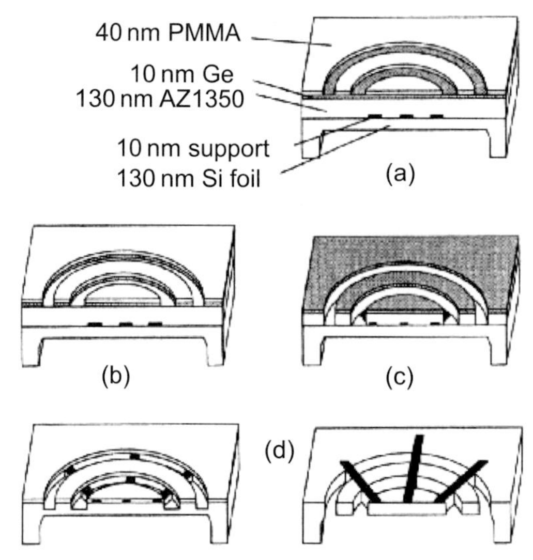

Figure 4.25 Generation of zone plates for HAS microscopy. (a) Illumination and development. (b) – (d) reactive ion etching with $CBrF_3$ (b, d) and O_2 (c). Reprinted with permission from [277]. Copyright 1999 Elsevier Science B.V.

A possibility is to use bent single crystalline surfaces, which allow focus diameters of 210 μm [272–274]. Appropriate single crystals are, for example, Si(111) crystals with 50 μm thickness. The crystals are attached as the second electrode of a metallic plate capacitor using 0.25 mm thick insulating spacer layers. If one applies an electric field of 10^7 $\mathrm{Vm^{-1}}$ between the plates, an electrostatic pressure

$$P_{es} = \frac{\epsilon_0 E^2}{2} \tag{4.12}$$

of about 440 $\mathrm{Nm^{-2}}$ is generated (about 4 mbar), which leads to an elastic deformation of the silicon crystal [275]. By use of an asymmetrically bent, elliptical silicon mirror it should become possible to correct imaging errors in the focus and thus one should be able to obtain focus diameters below one micrometer [276].

Another, and more flexible, possibility is to use zone plates in analogy to X-ray microscopy. In the case of matter beams, the rings of the plates have to be free standing, that is, there is no material between the rings and they are supported by vertical bars. Such zone plates can be produced with the appropriate structure size via reactive ion etching (Figure 4.25). First an etch mask is produced in the form of a 130 nm thick silicon foil, which contains the pattern of the perpendicular bars as 10 nm thick chromium bars. Onto this foil 130 nm dye, 10 nm germanium and

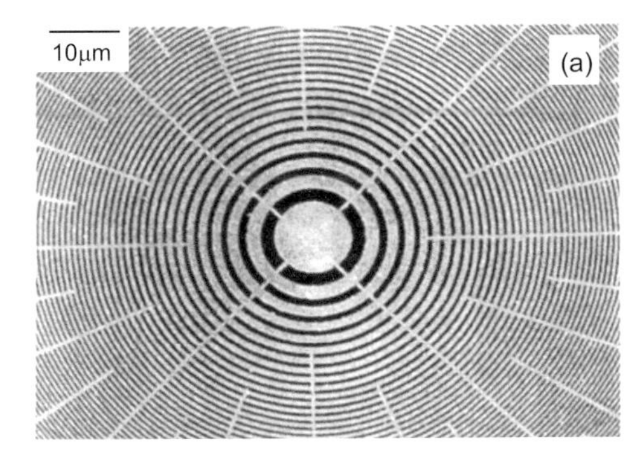

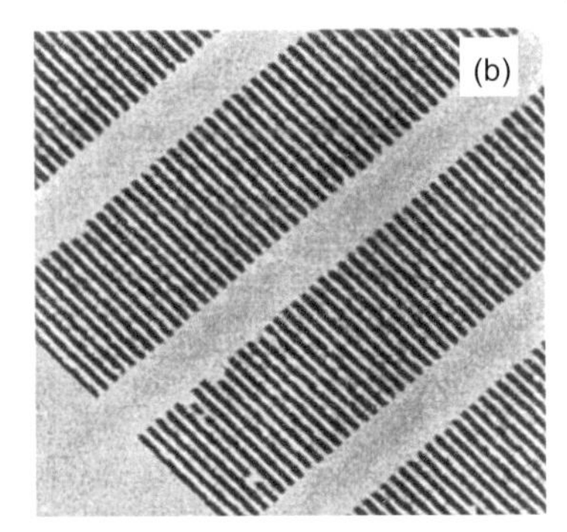

Figure 4.26 Scanning electron microscopy image of a free standing zone plate for helium microscopy. (a) central region, (b) outer region; the outer slits have widths of 50 nm. Reprinted with permission from [277]. Copyright 1999 Elsevier Science B.V.

40 nm PMMA layers are evaporated and illuminated with the zone plate pattern. After development the zone plate structures are reactively ion etched in a $CBrF_3$ plasma on the 10 nm germanium layer and thereafter in the dye layer. This dye mask is then used in a final etch step in $CBrF_3$ in order to structure the silicon foil with the zone plate pattern. The supporting bars made of chromium are not affected by the etching process.

Figure 4.26 shows the central and outer parts of such a zone plate with a diameter of 0.5 mm, obtained with an electron microscope. The width of the outermost ring of 50 nm corresponds to a possible resolution of about 60 nm.

As a first step towards a 'helium microscope' it has been demonstrated that the zone plate shown in Figure 4.26 is capable of focusing a helium beam from initially 400 μm to 2 μm (Figure 4.27). The helium beam has been generated using the method discussed in Section 4.4. However, special glass skimmers have been used with diameters of just a few microns. These extremely small skimmers result in a nearly point-like atom source (albeit with somewhat lowered intensity).

As compared to focusing attempts in the past using metastable atoms [278], the focussing discussed here means a demagnification of the spot diameter by a factor of ten and – even more important – an increase in the brilliance by a factor of 10^8. While the spatial resolution of this beam is 'just' of the order of a diffraction limited light focus, the high brilliance indicates that a high resolution helium microscope might become possible using more advanced atom optics or selective scattering processes. Helium atoms are definitely the most gentle particles for microscopic imaging.

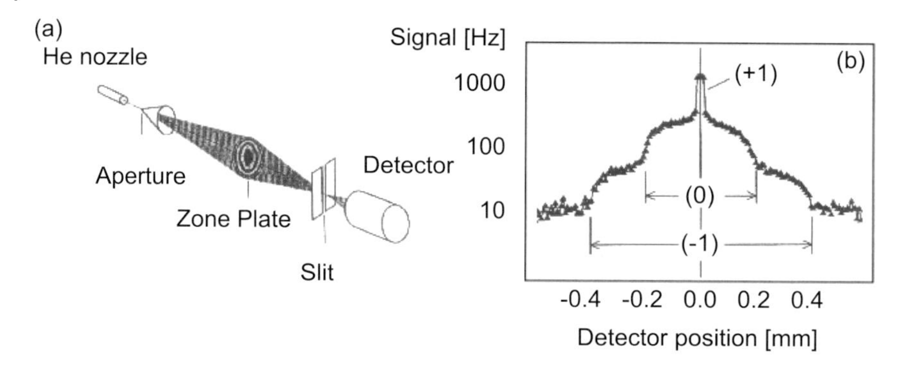

Figure 4.27 Schematic set up (a) and typical intensity distribution (b) of a helium beam diffracted at a zone plate. The numbers $(+1)$, (0) and (-1) represent focussed, non-diffracted and defocused intensity distributions. Reproduced with permission from [279]. Copyright 1999 American Physical Society.

An application for the diffraction of helium atom beams with immediate importance for nanotechnology is the characterization of transmission gratings with grating constants of the order of 100 nm. The diffraction pattern of a low energy helium atom beam from such a grating can be reproduced quantitatively only if the average slit width, the true profile of the rods (e.g. trapezoidal) and the statistical or periodic disorder of the lattice has been taken into account[9] [282]. All these factors can be also determined from a comparison of the results of a transmission measurement and a simulation.

It is finally noted that not only thermal atom beams are of interest for new microscopies, but also coherent atom beams which are generated, for example, from a Bose–Einstein condensate [283]. In principle this could result in a variety of new possibilities similar to the multitude of laser based microscopies that complement thermal light source microscopies. However, so far, the handling of coherent atom beams is far from simple enough to be applied to microscopic devices.

9) For heavy particles such as xenon the long range C_3/z^3 interaction potential between particle and lattice rods and the surface roughness of the rods also play an important role [280]. If these factors are known, one can deduce from the decrease of the effective slit width the size of the particles that have been transmitted through the grating. This led to a determination of bonding length r and binding energy E_b of the helium dimer ($r = 52 \pm 4$ Å; $E_b = 11$ mK) [281].

4.3 Linear and Nonlinear Spectroscopy

Optical spectroscopy is of importance for nanophysics and nanotechnology as it allows a destructionless measurement of modified electronic and optical properties of nanoscaled aggregates. An extensive and modern description of spectroscopic methods using coherent laser light can be found in [284]. All these methods can be applied without significant variations to nanoscaled systems. Of special interest for the investigation of *isolated* nano particles is the single molecule spectroscopy, which usually employs laser-induced fluorescence. Besides the properties of the particles themselves, they can also be used as sensors for changes in their environment. For example, if the particles are in an aqueous solution, the pH-value of the solution can be determined with high spatial resolution [285].

In what follows we discuss as an addendum to the classical spectroscopic methods the possibilities to characterize micro-nanoscaled structures using optical methods.

Linear Spectroscopy

Linear spectroscopy allows one to obtain important structural and morphological information about microscaled objects. In many cases the polarization of light and interference phenomena are employed.

The first example (Figure 4.28) shows how one can determine the thickness of micrometer thick, transparent layers via multiple beam interference .

The transmission $T(\lambda, \theta)$ of light through a homogeneous layer m of thickness L with index of refraction n_m, situated in between two homogeneous media '1' and '2' is given by [286]

$$T(\lambda, \theta) = \left| \frac{t_{1m}^{p,s} t_{m2}^{p,s} \exp(i\, w_m L)}{1 - r_{m1}^{p,s} r_{m2}^{p,s} \exp(2i\, w_m L)} \right|^2 \quad . \tag{4.13}$$

The Fresnel coefficients of both polarizations p and s for reflection $r_{ij}^{p,s}$ and transmission $t_{ij}^{p,s}$ at the interface between media 'i' and 'j' and the function w_m can be found, for example, in [288]. Equation 4.13 describes the transmission including all multiple beam interferences via the second sum in the denominator.

Figure 4.28 shows such thickness determination at normal incidence of light $\theta = 0°$ for the example of two wavelength regions: a) $525 - 545$ nm and b) $1050 - 1150$ nm. The points are measured, the solid lines are calculated transmission spectra following Equation 4.13 assuming a thick-

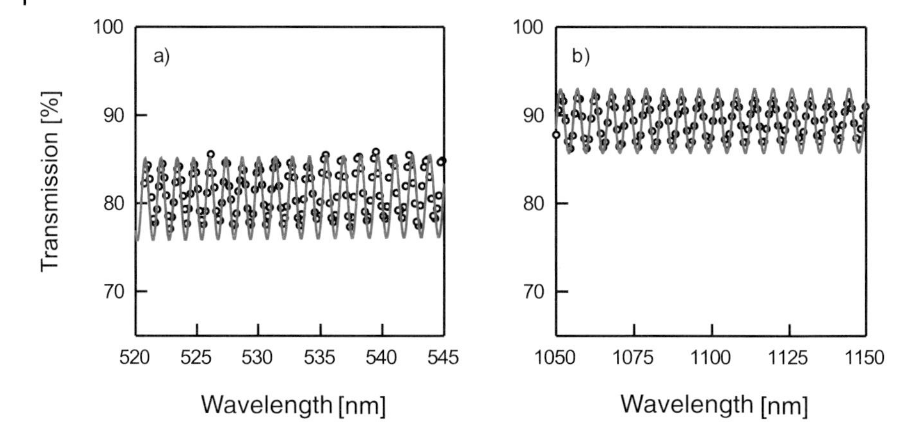

Figure 4.28 Photospectrometric thickness measurement: measured transmission (solid line) of a mica platelet as a function of wavelength for (a) 525–545 nm, and (b) 1050–1150 nm [287].

ness of the platelet of $L = 65.2\,\mu\text{m}$. In the following example it is shown that one can obtain via linear spectroscopy morphological information even for objects which are much smaller than the wavelength of the applied light. Here one uses the fact that light scattering at objects which are smaller than the wavelength of the scattering light depends nonlinearly on the size of the particles ('Mie theory', see Section 6.1.4). One finds for the total cross section, i.e. the probability for total extinction of light of wavelength σ by a dielectric sphere with radius a and dielectric function ϵ [289]

$$\sigma = \frac{8\pi}{3} k^4 a^6 \left| \frac{\epsilon - 1}{\epsilon + 2} \right|^2 \tag{4.14}$$

with wavevector $k = 2\pi/\lambda$.

For a metallic sphere with radius a, which is embedded in vacuum with a complex dielectric function $\epsilon = \epsilon_1 + i\epsilon_2$, one finds in the quasistationary limit ($a \ll \lambda$) for the total scattering cross section:

$$\sigma \propto \frac{a^3}{\lambda} \frac{\epsilon_2}{(\epsilon_1 + 2) + \epsilon_2} \quad . \tag{4.15}$$

For weakly absorbing materials ($\epsilon_2 \ll 1$) the cross section has a resonance at $\epsilon_1 = -2$.

In both cases the reflectivity of the spheres depends strongly on their size a as well as the wavelength of the impinging wavelength λ. If one measures extinction as a function of wavelength for particles of fixed

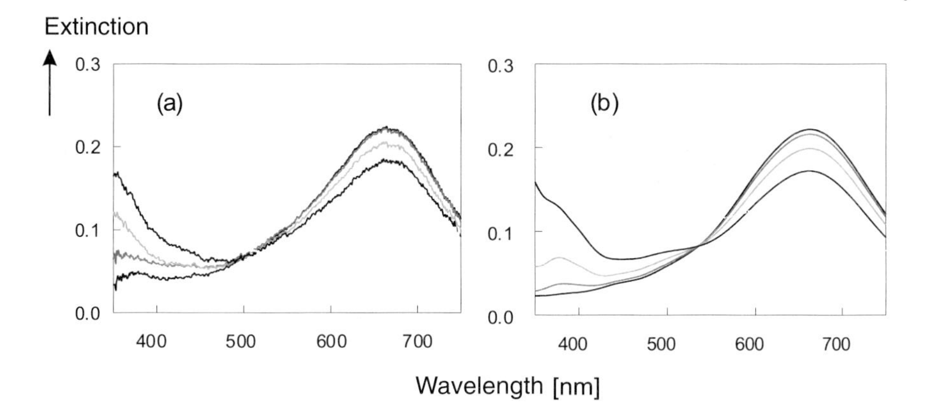

Figure 4.29 (a) Experimentally measured change in optical spectra of strongly oblate sodium ellipsoids ($< a >\approx 80\,\mathrm{nm}$, $R = c/a = 0.27$) as a function of polar angle Θ (surface temperature $T_S = 150\,\mathrm{K}$). The lowest curve is for $\theta = 0°$, the uppermost for $\theta = 60°$. (b) Theory.

size, one observes a dipole resonance at a wavelength that is specific for given size. This is demonstrated in Figure 4.29 for the example of alkali clusters, adsorbed on a dielectric. The theoretical spectra (Figure 4.29b), which fit the experimental data very well, use the known dielectric functions and the size distributions of the clusters – hence one can determine optically these size distributions. Apparently the clusters on surfaces are not spheres but ellipsoids, which posses besides the semiaxis a parallel to the surface a second one, c, which is oriented perpendicular to the surface. A suitable parameter for characterizing the ellipticity is thus the ratio of semiaxes, $R = c/a$.

The precision of this optical method can be evaluated by comparing the results with data from force microscopy measurements. In Figure 4.30 such a direct comparison has been made for oxidized alkali clusters on a mica surface [290]. The force microscopy results in size distributions (Figure 6.19), which can be used to calculate a theoretical extinction spectrum (grey curve in Figure 4.30). This curve agrees nearly perfectly with the curve from optical measurements – deviations being due to an non-precise determination of the absolute number of clusters on the surface.

Vice versa, a fit of the optical spectra allows one to determine distribution functions, which in turn can be compared with the distributions that have been measured via force microscopy. In Figure 6.19 it is shown that this comparison also reveals good agreement.

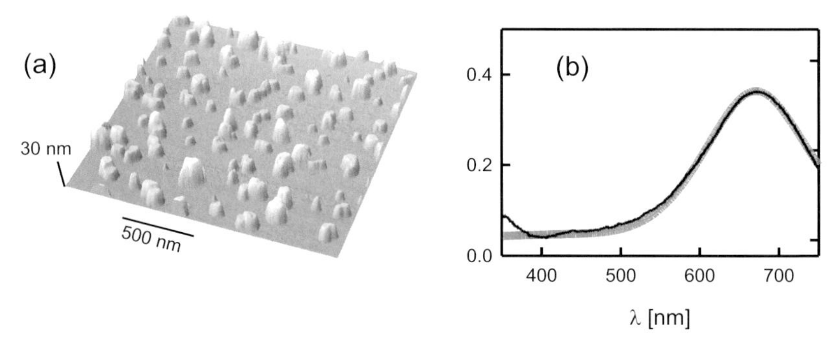

Figure 4.30 Precision of the optical analysis: (a) AFM image of an alkali cluster distribution on a mica surface. (b) Calculated (grey curve) using the function in (a) and measured (black curve) optical spectra.

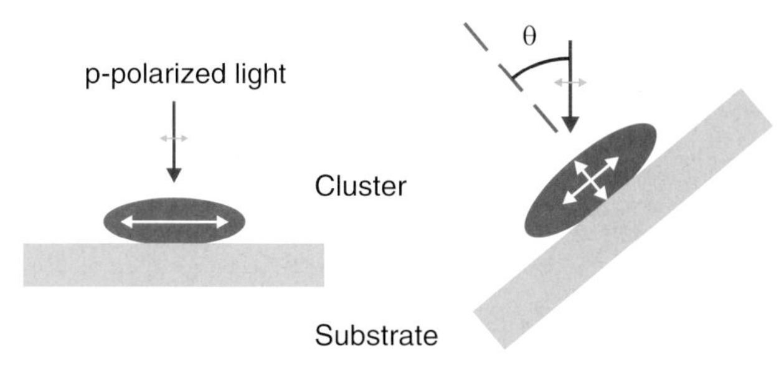

Figure 4.31 Sketch of p-polarized excitation of optically active ellipsoids on surfaces. For various values of the polar angle Θ the semiaxes of the ellipsoids are excited differently strong, for $\Theta = 0^\circ$ only the large semiaxis is excited.

The vectorial nature of light allows one to obtain detailed information about the ellipticity of nanoscaled particles. The principle is sketched in Figure 4.31. If one applies p-polarized light with an electric field vector that oscillates in the plane of incidence and if one changes the angle of incidence θ, for $\theta = 0^\circ$ the clusters can be excited solely along their semiaxis a parallel to the surface. In this case one expects to observe a single resonance in the wavelength dependence, corresponding to the average cluster size. For increasing θ the electric field vector also has a component along the semiaxis c perpendicular to the surface. Is the cluster an ellipsoid, this semiaxis is smaller (for oblate cluster) or larger (for prolate cluster) compared to a, resulting in the appearance of a second resonance. The relative intensity of this resonance should increase with increasing angle of incidence. In Figure 4.29 it is shown that this simple idea is indeed realized. The clusters on the surface are thus ellipsoids as sketched in Figure 4.31, and as verified in Figures 3.22 and 6.30 via molecular dynamic calculations and force microscopy.

Evanescent Waves and Surface Plasmons

An interesting method to focus the optical detection onto the nanoscaled surface aggregates or the surface layer is to implement resonant excitations in the surface film. Such resonant oscillations of the electron gas of density N_e with respect to the positively charged crystal lattice have the bulk plasmon frequency $\omega_p^2 = N_e e^2/(m_e \epsilon_0)$. The restoring force for the plasmon oscillation following excitation by an external field is the field that is generated by the extension itself. The quanta of this oscillation ('plasmons') possess energies $h\nu$ of the order of a few electron volts.

At the interface between a metal with the complex dielectric function $\epsilon_1 = \epsilon_1' + i \cdot \epsilon_1''$ and a gas or the vacuum with the constant ϵ_2, electromagnetic boundary conditions allow for p-polarized light a specific class of solutions for the governing wave equations (Maxwell's equation), namely the 'surface plasmon polaritons' [291]. They propagate along the surface with wavevector $k_x = 2\pi/\lambda_x$ and frequency dependence

$$k_x = \frac{\omega}{c}\sqrt{\frac{\epsilon_1 \epsilon_2}{\epsilon_1 + \epsilon_2}} \quad . \tag{4.16}$$

Along a smooth surface the surface plasmons are damped in the absorbing metal (imaginary part of the dielectric function) and thus decay with a characteristic length of a few micrometers. The energy of the plasmons is converted into Joule heating of the metal. In the case of a rough surface the plasmons can also decay via light radiation, decreasing the decay length significantly. In Figure 4.32 fluorescence images of such radiatively decaying surface plasmons are shown.

In order to obtain these images, a 70 nm thick silver film was adsorbed on a glass substrate and covered with a 10 nm thick silicon oxide layer as spacer layer for a small fraction of a monolayer laser dye. The laser dye contains the probe molecules to be excited by the light scattered or radiated by the plasmons. The plasmon excitation in the silver film is made possible via lithographically fabricated defects (silver nanoparticles of different shape), which are excited locally within the focus of a microscope objective using laser light. The decaying plasmons excite the dye molecules, which emit light that is spectrally shifted. This red shift allows one to detect the light nearly background-free. The $cos^2\Theta$-distribution of the plasmon radiation characteristics from a point defect is seen in Figure 4.32a [293]. A line defect (a silver wire of width 200 nm, height 60 nm and length 20 μm) generates an intensity distribution given

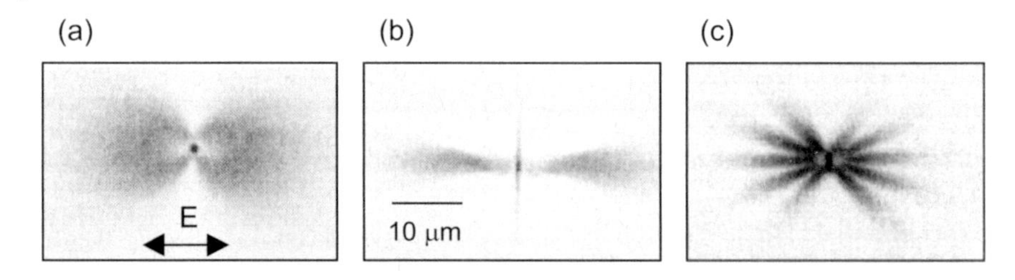

Figure 4.32 Images of radiating surface plasmons, excited at a silver nanoparticle with diameter 200 nm and height 60 nm (a), at a silver wire (b) and at two silver nanoparticles deposited at a mutual distance of 1.5 μm (c). The orientation of the exciting electric field vector is also plotted [292].

by the exciting focus diameter (Figure 4.32b), and two narrow spaced point defects result in an interference pattern (Figure 4.32c).

In all cases the plasmon oscillation is localized close to the excitation point and at the metal surface. Perpendicular to the surface (in z direction) the field amplitude decreases exponentially with the characteristic length (1/e depth)

$$z_i = \frac{\lambda}{2\pi}\sqrt{\frac{\epsilon_1' + \epsilon_2}{\epsilon_i'\epsilon_i'}} \quad , \quad i = 1,2 \quad . \tag{4.17}$$

For silver at a wavelength of 600 nm the decay length in the metal (index '1') is 24 nm, those into the surrounding vacuum (index '2') 390 nm.

Momentum conservation forbids the decay of surface plasmons on an ideally smooth surface. At the same instance, optical excitation of plasmons via photons is not possible. If one irradiates the surface with electrons, the change of momentum transfer conditions via change of scattering angle inside the solid allows one to fulfill momentum conservation and thus to excite plasmons. In order to couple light to surface plasmons, an increase of the wavevector by Δk_x is necessary for a given photon energy. This can be achieved via a lattice structure on the surface or more generally via surface roughness; such roughness adds additional reciprocal lattice vectors to the wavevector of the incoming light.

If one evaporates the metal film, inside which the plasmons are to be excited, onto a prism one can generate via total internal reflection inside the prism an evanescent wave, which is phase matched to the surface plasmon wave under a certain angle of incidence. This angle is determined by the dielectric function of the film (Kretschmann–Raether configuration [294]). The evanescent wave at the point where the irradiating light hits the hypothenuse from the prism side propagates over the sur-

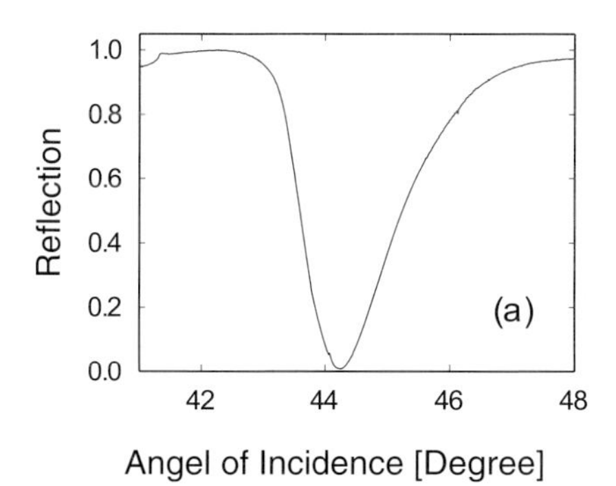

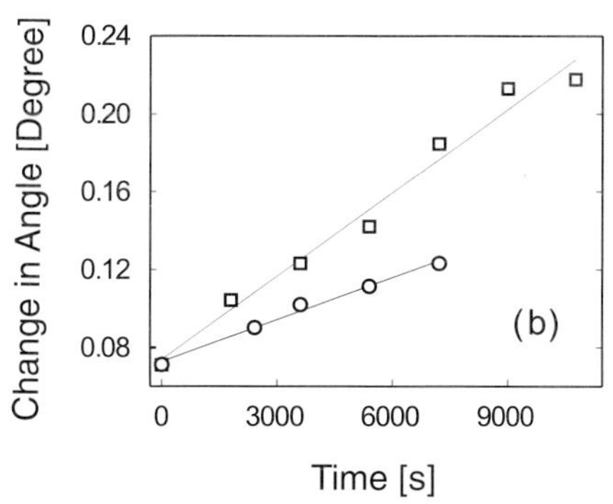

Figure 4.33 (a) ATR minimum, measured at the interface gold/vacuum using p-polarized HeNe laser light. The gold film (25 nm thickness) has been evaporated onto a quartz prism. The angle of incidence is the angle, which the light inside the prism has with respect to the normal on the hypothenuse. (b) Measured shift of the ATR minimum for a water film (squares) and an NO film (circles) of increasing thickness [295].

face without energy transport. Their amplitude is exponentially damped normal to the prism surface. If a thin film is evaporated onto the prism, energy can be dissipated which subsequently is missing in the reflected light intensity. If one modifies the angle of incidence of the light onto the prism, one obtains a characteristic minimum in the reflectivity ('attenuated total reflection', ATR). In Figure 4.33 an example is shown for the reflected intensity of a HeNe laser (632.8 nm) on a thin gold film. If one grows aggregates on the gold film, the position of the ATR minimum shifts. A dielectric film of 100 nm thickness results in a shift of about 0.16 degrees.

The depth of the minimum, that is, the efficiency of coupling energy into the film, depends on the film thickness and the film roughness since those factors affect damping of the incoming and reflected waves, as well as the phase relationship of the partial waves that are generated at the prism/metal and the metal/vacuum interfaces. For a silver film and an excitation wavelength of 500 nm one obtains minimum reflectivity for a thickness of 55 nm. This ATR minimum relates to an enhancement of the electromagnetic field at the metal/vacuum interface of up to two orders of magnitude. Of course, such enhancement is advantageous for nonlinear optical spectroscopies or new, extremely sensitive nanocomponents. That way, for example, a photonic transistor can be fabricated on the basis of local plasmon excitation [296].

Since the ATR method allows as an alternative to ellipsometry [297] a quantitative determination of the dielectric function of evaporated thin films, [298] and since it reacts very sensitive to changes of these films via, for example, adsorbates, a multitude of optical sensors in technological and medical areas have been developed that base on total internal reflection. The advantage for the application in the medical-biological field as biosensors [299] is that labelled or labelling molecules are not necessary. Rather one can adsorb bioreceptors with specific sensitivity to the substance to be investigated (antibodies, for example) on the sensor surface and so measure with high sensitivity for example how much, of a specific substance is binding from the liquid phase onto these receptors (cf. Figure 4.33b).

Even without damping via an evaporated, absorbing film and following plasmon excitation the evanescent character of lightwaves at the interface between two media is currently used for a large multitude of surface characteristic and nanooptic methods. Examples include near field optics, guiding of atoms along nanostructured interfaces or surface sensitive spectroscopies such as 'selective reflection' [300] or multiphoton evanescent-volume wave spectroscopy [301,302]. See also [303] and especially [304,305] for an introduction to the use of evanescent waves.

Nonlinear Spectroscopy

Detailed information about nanoscaled objects can be obtained optically via nonlinear spectroscopies. Due to the high photon densities that laser light provides, nonlinear effects can be obtained easily and in numerous forms once a laser beam hits a solid or fluid body.

Optical Frequency Doubling

The electromagnetic field $\vec{E}(\omega)$ of a light wave of frequency ω, that hits a polarizable medium, induces electron oscillations, which sum up to a macroscopic polarization $\vec{P}(\omega)$. The radiation of the excited electrons results in reflection or diffraction of light. If the electrons oscillate in the potential field of a harmonic oscillator, the frequency of the emitted light is not changed relative to the frequency of the irradiating light,that is, $\vec{P}(\omega) \propto \chi\vec{E}(\omega)$. The material dependent proportionality constant χ, which determines intensity and phase of the reflected field, is called 'susceptibility'.

With increasing field strength the electron motion becomes anharmonic: in the emitted wave one observes in addition to the fundamental frequency ω multiples 2ω, 3ω and so on. This is at least the case when the external field strength is of comparable size to the intraatomic field strength. The electrostatic field that acts on a ground state electron in the hydrogen atom has, for example, a strength of $5.14 \cdot 10^9\ \mathrm{Vcm}^{-1}$. Such a field strength can be reached in the 10 μm focus of a laser with 10 ns pulse duration and a pulse energy of 0.2 J.

The induced polarization can be written as a Taylor expansion with respect to the electric field [289] :

$$\vec{P} = \epsilon_0[\chi^{(1)}\vec{E} + \chi^{(2)}\vec{E}\vec{E} + \chi^{(3)}\vec{E}\vec{E}\vec{E} + ...] \tag{4.18}$$

with dielectric constant of the vacuum, $\epsilon_0 = 8.86 \cdot 10^{-14}\,\mathrm{AsV^{-1}} \cdot \mathrm{cm}$. Instead of the proportionality constant χ one finds as coupling elements between light and matter the susceptibility tensors $\chi^{(i)}$, which describe the anisotropic, nonlinear properties of the medium. Typically $\chi^{(2)}$ is about ten orders of magnitude smaller than $\chi^{(1)}$, that is, $\approx 10^{-10}\,\mathrm{cmV^{-1}}$ and $\chi^{(3)} \approx 10^{-17}\,\mathrm{cm^2V^{-2}}$.

For the determination of characteristic electronic or morphologic properties of nanostructures the interface sensitivity of nonlinear optics is most important. The reason is that at the interface the inversion symmetry of the infinitely extended bulk is broken. This loss of inversion symmetry in turn is important for the generation of a nonlinear polarisation of *even* order inside the nonlinear medium. If the medium has inversion symmetry, the even electric multipole terms (dipole, octupole ...) in the multipole expansion of the electromagnetic fields vanish since the fields $\vec{E}$ change sign upon parity operation, $\vec{P}(-\vec{E}) = -\vec{P}(\vec{E})$.

In media with inversion symmetry (fcc and bcc metals, silicon, gases and liquids or glasses) one finds thus in the bulk only nonlinear terms of odd order (from $\chi^{(3)}$). If one observes frequency doubling, it results from surface contributions (in dipole approximation) and thus contains surface characteristic information. Since the main reason for the nonlinear polarization is the generation of a strong dipole field at the surface, this polarization will be localized in the uppermost atomic layers down to a depth of one to a few nanometers [286].

The total susceptibility of surface adsorbed structures is composed from the nonlinear polarizability of the individual aggregate, $\alpha^{(2)}$, and the number density of surface aggregates, N_S,

$$\chi_S^{(2)} = N_S \cdot \alpha^{(2)} \quad , \tag{4.19}$$

if one can average over the orientation of the aggregates and can neglect their mutual interactions falls. An enhancement factor for this susceptibility, which is especially important for nanostructures, is the local field strength tensor **L** at the surface,

$$\chi_S^{(2)} \rightarrow \mathbf{L}(2\omega)\chi_S^{(2)}\mathbf{L}(\omega)\mathbf{L}(\omega) \quad . \tag{4.20}$$

In the case of an ideal smooth surface L is given by the linear optical Fresnel factors $f_i(\omega, 2\omega)$. Explicite terms for ω and 2ω can be found, for example, in [288]. These factors depend essentially on the angle of incidence and average over the optical properties down to the optical penetration depth δ of the light. The penetration depth is $\delta = \lambda/2\pi Im(n)$, where $Im(n)$ describes the imaginary part of the index of refraction, which governs absorption. For a rough surface electromagnetic field enhancement can be very large, especially if morphology dependent collective resonances ('surface plasmon excitations') are observed.

Optical frequency doubling allows one to determine electronic symmetries of surface structures by taking advantage of the vector character of electromagnetic radiation. The $\chi^{(2)}$ tensor reflects the symmetries of the surface via the symmetry properties of the electronic surface states. Since it is a tensor of rank three, only three- or lower-fold surface symmetries can be resolved.[10]

The second order nonlinear component of Equation 4.18 is given by:

$$P_l(2\omega) = \sum_{m,n} \chi^{(2)}_{lmn}(-2\omega;\omega,\omega)E_m(\omega)E_n(\omega), \tag{4.21}$$

where $\chi^{(2)}_{lmn} = |\chi^{(2)}_{lmn}|e^{i\phi_{lmn}}$ couples the mth and nth components of the fundamental with the lth components of the generated nonlinear polarization (ϕ is the phase). The $\chi^{(2)}$ tensor, due to the condition $\omega_1 = \omega_2 = \omega$, might be contracted to 18 components ('piezoelectric contraction'), which – depending on the symmetry of the surface – are not independent of each other [307]. If the exciting laser irradiates along the surface normal (z-direction), then the emitted SH intensity is the following function of the angle Θ between the electric field vector of the linearly polarized laser and preferred directions on the crystal (labelled x and y) [308],

$$I_x(2\omega) \propto |(\chi^{(2)}_S)_{\|\|\|}\cos^2\Theta + (\chi^{(2)}_S)_{\|\perp\perp}\sin^2\Theta + (\chi^{(2)}_S)_{\|\perp\|}\sin2\Theta|^2 \tag{4.22}$$

and

$$I_y(2\omega) \propto |(\chi^{(2)}_S)_{\perp\|\|}\cos^2\Theta + (\chi^{(2)}_S)_{\perp\perp\perp}\sin^2\Theta + (\chi^{(2)}_S)_{\perp\|\perp}\sin2\Theta|^2. \tag{4.23}$$

In the case of a medium that is symmetric in the surface plane (4 mm symmetry) only three independent components exist: $d_{33} = 2\cdot\chi_{\perp\perp\perp}, d_{31} = 2\cdot\chi_{\perp\|\|}$ and $d_{15} = 2\cdot\chi_{\|\|\perp}$.

In order to determine these independent components, SH measurements are performed under selected polarization combinations; that is, the polarization direction of the incoming laser beam is selected to be

10) At least in dipole approximation ($L = 1$). If it is possible to measure higher order multipole-contributions, then the highest resolvable rotational symmetry is for an Nth order nonlinear technique given by $(N + L)$ [306].

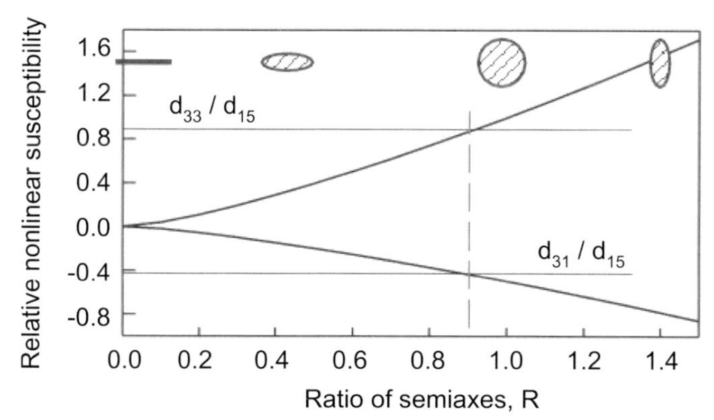

Figure 4.34 Variation of the ratio between components of the nonlinear susceptibility tensor as a function of ellipticity of the nano particles. The dashed perpendicular line connects measured values of the relative nonlinear susceptibilities.

parallel ('p-polarized') or perpendicular ('s-polarized') with respect to the plane of incidence; another choice is 45°. The SH signal can also be observed under well defined polarization directions.

From a measurement of the SH intensities for polarization combinations sp and $45°s$ one obtains the ratio of the tensor components $|d_{31}|/|d_{15}|$ and can thus deduce from the measurement in the pp configuration the d_{33} component. In order to conclude the average particle shape from the measured tensor components a 'projection model' might be applied [309,310]. This assumes that the particles are much smaller than the wavelength of the light involved, that is, much smaller than a few hundred nanometers. One obtains for a spheroid with the axial ratio $R = a/b$ with a the large and b the small semiaxis, a tensor with the components

$$\begin{aligned} \chi_{\perp\perp\perp} &= \frac{1}{2}\chi_0\, C(R) \\ \chi_{\perp\|\|} &= -\frac{1}{4}\chi_0\, C(R) \\ \chi_{\|\|\perp} &= \frac{1}{2}\chi_0\, [2 - C(R)] \end{aligned} \tag{4.24}$$

Here, χ_0 means the intrinsic nonlinear susceptibility per unit bent surface of the particle, and the function $C(R)$ depends on the semiaxis ratio as

$$C(R) = 2R^2 \frac{R^2 - 1 - 2\ln R}{(R^2-1)^2} \quad . \tag{4.25}$$

In Figure 4.34 the ratios d_{33}/d_{15} and d_{31}/d_{15} are plotted as a function of deformation R. For a plane interface ($R = 0$) both the ratios d_{33}/d_{15}

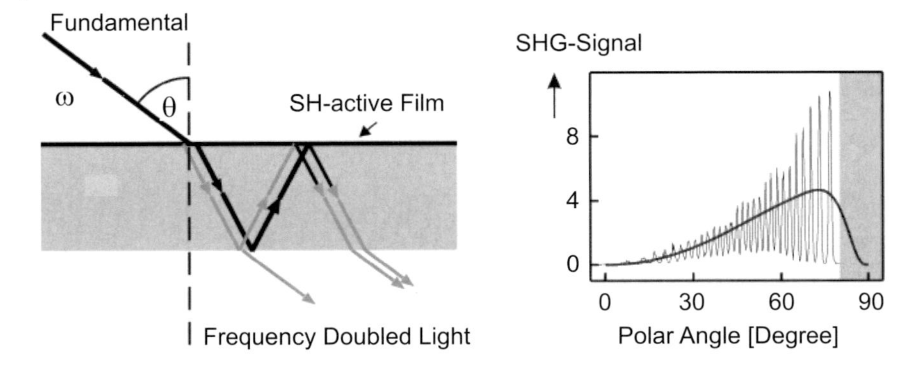

Figure 4.35 Frequency doubling at a thin surface layer and subsequent multiple beam interference. On the right-hand side a measured polar angle dependence is shown from metal clusters supported on a thin mica substrate. The incoming light has been s-polarized, the detected frequency doubled light p-polarized.

and d_{31}/d_{15} vanish. With increasing R the values increase and reach the limits for spherical particles, which are 1 and $-1/2$. The perpendicular, dashed line connects the measured values for sodium clusters on a lithium fluoride surface at room temperature [287]. It is obvious that the clusters in this specific case are somewhat deformed spheroids.

The whole story becomes more complicated if the nanostructures are adsorbed on very thin substrates (thin compared to the exciting wavelength), since reflections at the interfaces lead to interference effects. Figure 4.35 demonstrates this fact with the help of an experimentally observed change of SH intensity as a function of the polar angle for an optically active alkali film, adsorbed onto a thin mica plate [311]. The solid curve results from a calculation using incoherent addition of SH light from both the front and the rear. Obviously the measured dependence can be reproduced only for sufficiently thick substrates. If one takes into account rear side reflection and if one adds the corresponding contributions coherently, the positions of calculated and measured maxima are in very good agreement (curve 3 in Figure 4.36; curve 1 is the expected angular dependence if one does not take rear side reflection into account).

Further modifications of the SH intensity as a function of angle of incidence appear if the signal has contributions from spatially separated sources. These could be, for example, different nonlinearly optically active ultrathin films, which again result in oscillations as a function of angle of incidence.

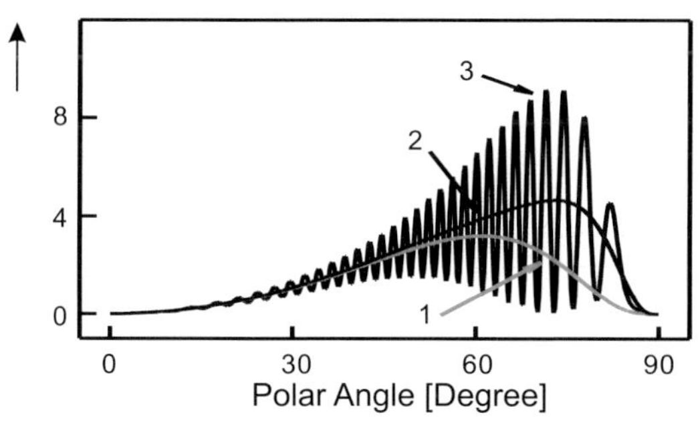

Figure 4.36 Calculated polar angle dependence of the frequency doubled light for the conditions of Figure 4.35. The curves '1' and '2' result from calculations assuming noncoherent light without (1) and with (2) back side source, the curve '3' is the complete, coherent calculation.

These oscillations can be explained by the dispersion of the substrate, which results in a phase difference between frequency doubled light from the front source (induced via the impinging fundamental light source with λ_ω) and SH from the rear side (induced via the fundamental, which penetrates the substrate). A comparison of measured and calculated curves results thus in information about the morphology of nanoscaled aggregates even at hidden interfaces. Such information is not available from direct structural investigations via, for example, scanning tunneling microscopy.

It is worthwhile to note again that the characteristic structure sizes that have been deduced via these optical methods are much smaller than the light wave applied. The small focal diameter of diffraction limited light (a few μm) and the corresponding good spatial resolution even allow nonlinear surface microscopy via SH generation, for example for the imaging of semiconductor quantum dots [258]. A huge advantage here is that an improvement of focus diameter and thus an improvement of resolution goes hand in hand with an increase in photon density and thus a strong increase in signal intensity. One gains in signal with increasing resolution. If in addition applies one ultrashort laser pulses, damage of the sample via thermal effects can be minimized.

Degenerate Four- Wave Mixing

Optical second harmonic and also, for example, sum frequency generation are three-wave mixing processes. The nonlinear polarization, which is necessary for the generation of the third photon, can be described by the second term in Equation (4.18). If one takes additional photons into

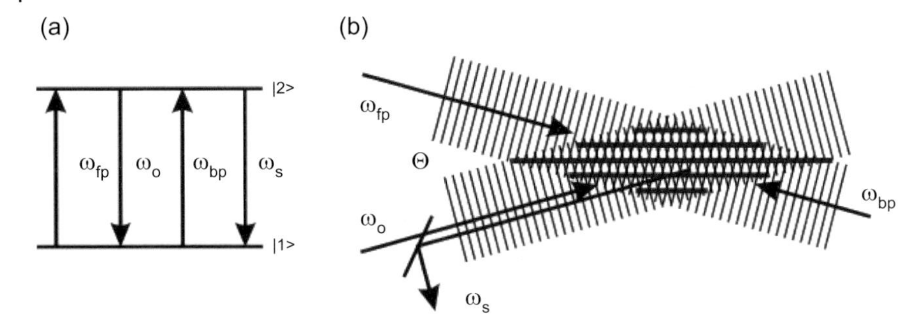

Figure 4.37 Energy and momentum conservation for degenerate four-wave mixing (DFWM). (a) Term scheme, (b) spatial superposition of two plane waves ω_{fp} and ω_o and read-out of the resulting lattice with a wave ω_{bp}. The wave vectors of the three superposed waves determine the direction of the signal wave ω_s.

account, multi-wave mixing processes and especially four-wave mixing takes place. The coupling constant between irradiating fields and induced polarization then is the nonlinear susceptibility of third order, $\chi^{(3)}$. Important four-wave mixing processes, which have been deduced by invoking energy conservation, are: coherent anti-Stokes Raman scattering (CARS) , frequency tripling (THG) and degenerate four-wave mixing (DFWM). In the latter case all photons have the same frequencies.

The term scheme for DFWM is especially simple: all four photons induce transitions between real states. This 'resonance enhancement' results in a higher signal strength compared with, for example, CARS. Momentum conservation determines the direction of the resulting signal wave. Since the phase-matching condition is:

$$\vec{k}_{\mathrm{fp}} + \vec{k}_{\mathrm{bp}} - \vec{k}_{\mathrm{p}} - \vec{k}_{\mathrm{s}} = 0, \tag{4.26}$$

$\vec{k}_{\mathrm{fp}} + \vec{k}_{\mathrm{bp}}$=0 and the energies of all involved photons are equal, the phase conjugate signal wave will counterpropagate to the probe wave (phase-conjugate (PC) geometry).

DFWM [312] is a real-time variant of optical holography, which has been known since the 1960s: two laser beams (here, forward pump and object beam) are overlapped coherently under a small angle Θ (Figure 4.37b). They induce in a nonlinear medium with susceptibility $\chi^{(3)}$ an interference pattern (grating), which contains information about the amplitude and phase relations between the contributing waves. This information is recovered via Bragg scattering using a third beam (backward pump) and generates a phase conjugate signal wave. In contrast to conventional holography, in the case of DFWM generation and recovery processes occur simultaneously. This has big advantages, for example, for the time-resolved investigation of electronic relaxation processes in ultrathin films (see Section 6.5).

'Phase conjugation' in this context means that the signal wave has the same wavefronts and phase relations as the object wave. Only the sign of the wavevector $\vec{k}$ has changed. This can be seen by summing over the irradiating waves. The electric wave of the forward pump is:

$$\vec{E}_{fp} \propto exp(i\vec{k}_{fp}\vec{r} - i\omega_{fp}t) \quad ; \tag{4.27}$$

that of the backward pump:

$$\vec{E}_{bp} \propto exp(i\vec{k}_{bp}\vec{r} - i\omega_{bp}t) \quad ; \tag{4.28}$$

and that of the object beam:

$$\vec{E}_{o} \propto exp(i\vec{k}_{o}\vec{r} - i\omega_{o}t) \quad . \tag{4.29}$$

Hence it follows for the signal:

$$\begin{aligned} \vec{E}_s &\propto exp(i(\vec{k}_s + \vec{k}_{fp} + \vec{k}_{bp})\vec{r} - i(\omega_s + \omega_{fp} + \omega_{bp})t) \\ &= exp(i\vec{k}_s\vec{r} + i\omega_s t), \end{aligned} \tag{4.30}$$

since $\vec{k}_{fp} = -\vec{k}_{bp}$ and $\vec{k}_o = -\vec{k}_s$ and $\omega_{fp} + \omega_o = 0$.

For example, the initial light wave might have passed through a phase-disturbing medium (e.g. an adsorbate) onto the nonlinear medium (e.g. a surface film). As long as the irradiated signal wave travels through the same phase-disturbing medium, the information content is not decreased.

Since all contributing photons are attributed to transitions between real states $|1>$ and $|2>$ (Figure 4.37a) a high intensity of the third order nonlinear signal follows ('resonance'). In such a case involving population transfer the generated grating might be a density grating. But even without strong absorption the coherent superposition of the laser beams results in a modulation of the complex index of refraction of the medium (amplitude- or polarization-grating), which leads to the generation of a phase-conjugate signal wave. The signal intensity is a measure of the depth of modulation of the grating just as in the case of diffraction of an external laser beam from the transient grating structure. One might name the whole process also 'transient grating scattering'. The lattice constant Λ (and thus the number of lattice rods within the overlap volume of the laser beams, i.e, the sensitivity of the method) depends on the crossing angle Θ via $\Lambda = \frac{\lambda}{2\sin(\Theta/2)}$. Thus a smaller crossing angle results in a more sensitive optical detection.

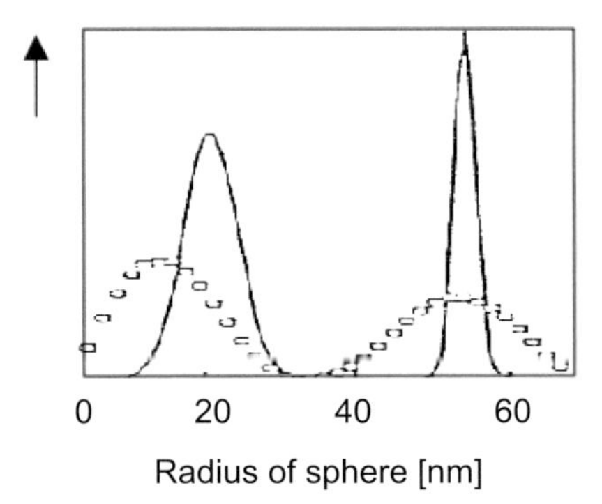

Figure 4.38 Size distribution of nanoscaled latex beads, determined via electron microscopy (solid line) and via optical correlation spectroscopy (squares) [313].

Correlation Spectroscopy

As another, and the final example, for the optical determination of morphological constants of nanoscaled particles we mention the homodyne correlation spectroscopy. Here, the frequency spectrum of the light that is scattered at the particle distribution (e.g. in a liquid) is measured in order to obtain information about the spatial distribution (and the size distribution) of the particles. The intensity distribution of the scattered light is evaluated from the autocorrelation spectrum, that is, from the frequency spectrum that has been determined with a photomultiplier. The intensity distribution depends nonlinearly on the size distribution of the scattering particles (cf. Equation 4.14), which allows one to deduce the size distribution with high precision. In Figure 4.38 such a measured distribution is compared directly to a distribution determined via electron microscopy, showing the validity of this approach.

4.4
Diffraction Methods

Diffraction of X-ray or neutron beams on solids results in information about the bulk lattice structure. Analogous information about the lattice structure of clean surfaces or of nanoscaled structures on surfaces can be obtained via surface scattering of nonpenetrating beams. Possible candidates for such beams are electrons, ions or neutral particles such as helium atoms. The local structure of surfaces or aggregates on surfaces can only be investigated if that structure has periodicity.

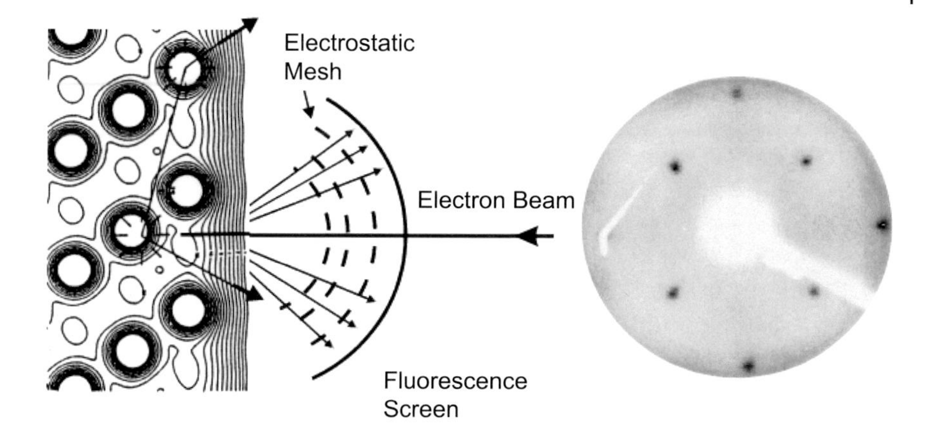

Figure 4.39 Electron diffraction on surfaces exemplified by a lithium fluoride surface. The open circles are symbols for the atom positions while the lines represent cuts through the planes of equal electron density. On the right-hand side the resulting LEED image is shown for an electron energy of 224.000×10^{-19} J (140 eV)

An important prerequisite for a Bragg diffraction pattern is that the de Broglie wavelength (Equation 2.1) $\lambda_{\mathrm{dB}} = h/|\vec{p}| = h/\sqrt{2mE}$ of the scattered beam is of the order of the lattice constant of the surface to be investigated. For electrons one finds

$$\lambda_{\mathrm{dB}}[\mathrm{nm}] = \sqrt{\frac{1.5}{E[\mathrm{eV}]}}, \tag{4.31}$$

that is, $\lambda_{\mathrm{dB}} = 0.12$ nm for 160.000×10^{-19} J (100 eV) electrons. Compared with typical lattice constants (Cu(100): 0.361 nm, mica(0001): 0.52 nm) one observes immediately that electrons are well suited for obtaining structural information from diffraction experiments.

Helium atom beams with the de Broglie wavelength

$$\lambda_{\mathrm{dB}}[\mathrm{nm}] = \sqrt{\frac{0.204}{E[\mathrm{meV}]}} \tag{4.32}$$

of approximately 0.1 nm at a speed of 920 ms^{-1} fulfill this condition.

The penetration depth of low energy electrons of a few tens to one hundred electron volts is of the order of 0.5–1 nm ('low energy electron diffraction', LEED [314]). LEED is thus sensitive to structural order in the uppermost two to three monolayers, including possible adsorbates. In Figure 4.39 a typical LEED set up is shown together with resulting diffraction image from lithium fluoride surface. The low energy electron beam is generated by an electron gun and is diffracted at the crystal, which is mounted on a movable manipulator inside an ultrahigh vacuum chamber.

The diffracted beams are sampled on a fluorescent screen, where they generate a picture of the reciprocal surface lattice. Intensity, position and width of the fluorescent points are detected with a video camera. In order to avoid inelastically diffracted electrons, a retarding electric field is applied between the crystal and the fluorescent screen.

From the number and position of the diffraction maxima one can deduce within the 'transfer width' of the instrument (meaning the spatial range of coherent scattering) the period of the surface lattice or its possible disturbance by adsorbate superlattices. For the LEED image in Figure 4.39, for the clean lithium fluoride surface one deduces a lattice constant of 0.403 nm. Typical transfer widths are from 10 nm for a conventional LEED up to 100 nm for a SPALEED, 'spot profile analysis LEED'. The intensity of the diffraction maxima contains information about the structure of the unit cell and the dynamics of the surface (e.g. the Debye–Waller factor $\exp(-\vec{K}^2 \langle u^2 \rangle)$ with scattering vector $\vec{K}$ and mean deflection $\langle u^2 \rangle$ of the thermally excited vibrations of the surface atoms).

An interesting extension of classical LEED is the LEEM (low-energy electron microscope), which allows one to detect surface particles with minimum sizes of 10 nm. At the same time, the movement of the particles can be observed [315, 316]. A LEEM is similar to a classical microscope: the illumination source is an electron source with an electrostatic lens, which produces an electron beam of 24.000×10^{-16} J (15 keV). The beam is deflected behind a sampling lens with a magnetic 90° deflector field. It is imaged onto the sample with an objective lens, which is kept at a potential of 15 kV: this decelerates the beam down to a few electron Volts. Because of the potential difference between sample and objective lens the low energy electrons, which are scattered from the sample, are accelerated to 24.000×10^{-16} J (15 keV) and are deflected in the opposite direction. They are imaged via an ocular lens onto a fluorescence screen.

If one uses only the specularly reflected beam by application of an aperture into the imaging optics, one obtains a bright field image of the sample surface with a lateral resolution of about 5 nm. The vertical resolution can be much better (a fews tenths of a nanometer) due to surface interference effects on structures with different heights. If, however, one selects one of the diffraction spots, one obtains an image of the periodic surface structures, which have generated this spot – this can be, for ex-

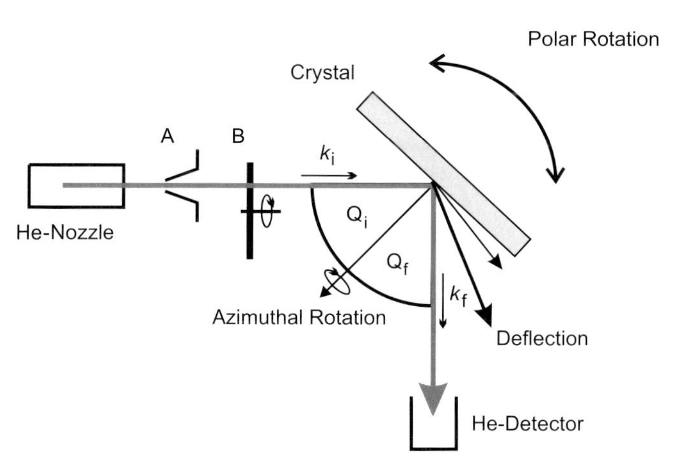

Figure 4.40 HAS set-up for surface diffraction of low-energy atoms. The nearly monochromatic helium atom beam is generated by expansion of helium gas under high pressure through a nozzle with a small orifice. The beam is skimmed via a specially shaped aperture ('skimmer') (A), chopped for energy analysis (B) and diffracted at the surface of a single crystal. The crystal can be rotated azimuthally around the surface normal and polar perpendicular to the surface normal.

ample, adsorbates, which can be selectively detected that way. Another interesting application is that one is able to obtain a LEED diffraction pattern from areas of the surface which have diameters of the order of a fraction of a micrometer. In contrast to conventional LEED, which uses an electron beam with a diameter of one to two millimeters, LEEM allows one to determine the different periodicities of neighboring domains or of nanoscaled aggregates on the surface.

The dispersion curves of surface phonons can be determined, and thus the dynamics of lattice vibrations can be investigated, via electron energy loss spectroscopy (EELS) [317]. The sample is irradiated by monoenergetic electrons with energies up to a few hundred electronvolts and the elastically and inelastically scattered electrons are analyzed with an energy resolution down to a few meV. The interpretation of the data is difficult since the penetration depth of the electrons results in contributions to the signal intensity not only from the topmost but also from lower lying atom layers. Since the cross section for photon excitation by electron collisions depends on the incidence energy and scattering geometry (meaning the surface projection of the involved momenta), a variation of these parameters in some cases allows one to discriminate between neighboring phonon dispersion curves, in spite of the relatively low energy resolution of the method. Applications include investigations of metals and chemisorbed adsorbates. However, in the case of insulators severe problems arise due to surface charging effects.

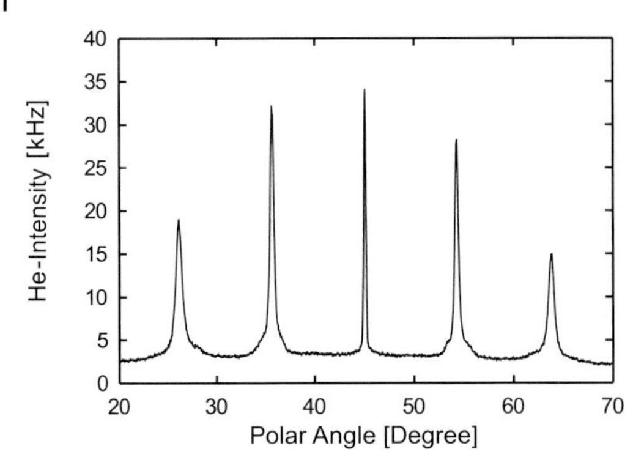

Figure 4.41 HAS angular distribution at an imperfect mica surface in $[11\bar{2}]$ direction. Incoming wave vector of the helium beam $\vec{k}_i = 6.4\,\text{Å}^{-1}$, surface temperature 245 K.

These problems can be avoided by using helium atom beams ('helium atom beam scattering', HAS) [318, 319] (Figure 4.40) . The reflected intensity distribution of an intense, monochromatic beam of ^{4}He atoms is detected by a magnetic mass spectrometer as a function of scattering angle with respect to the surface normal. The beam is generated by expanding helium gas under high pressure (≈ 100 bar) through an orifice of small diameter ($\approx 10\,\mu$m) at low temperatures (≈ 100 K) into a vacuum chamber [320]. During that expansion the non-directed thermal energy is transformed into a directed translational motion: a supersonic nozzle beam is generated. The beam with wavevector $\vec{k}_i$ is scattered at the single crystalline surface. One obtains information on the electronic density distribution about 0.4 nm *above* the position of the ionic cores of the crystal lattice. At this position the interaction between the electrons of the He atoms and that of the substrate result in a classical point of inversion. Hence HAS is a truly surface-sensitive method, which is well suited for the investigation of growth and change of adsorbate coverage.

In Figure 4.41 a measured angular distribution following diffraction at a single crystalline mica surface is plotted. Constructive interference is obtained for

$$\vec{K}_f - \vec{K}_i = \vec{G} + \vec{Q} \quad , \tag{4.33}$$

where $\vec{G}$ is a reciprocal lattice vector and $\vec{Q} = 2\pi/\lambda$ is a phonon vector[11]. This 'Laue–Bragg' equation follows immediately from momentum

11) In this equation we use $\vec{K} = \vec{k} sin\theta$ with θ the angle of incidence with respect to the surface normal instead of $\vec{k}$, since the parallel component of the wavevectors along the surface has to be taken into account for scattering on a two-dimensional lattice.

conservation for scattering with a surface. Energy conservation results in

$$(k_f^2 - k_i^2)\frac{\hbar^2}{2m} = \hbar\omega_Q, \qquad (4.34)$$

where $\hbar\omega_Q$ is the energy of an excited lattice vibration ('photon energy'). In the case of elastic scattering: $\vec{Q} = 0$, hence from the distance between diffraction maxima follows immediately the lattice constant of the investigated crystal surface. For Figure 4.41, $a = 0.52\,\text{nm}$ as lattice constant of the hexagonal unit cell of mica, analogous to LEED measurements.

Similar to LEED the small de Broglie wavelength of HAS results in a high structural resolution. The low kinetic energy (meV) and the monochromasy of the beam enable one to determine precisely the energy distribution of the scattered atoms via time-of-flight (TOF) measurements. The TOF spectra show distinct maxima for a given energy loss or gain, which are due to the excitation or annihilation of phonons [321]. Dispersion curves can be determined as a function of scattering angle. Due to the closed electronic shell of the ^{4}He atoms the attractive van der Waals forces are weak; hence the atoms are – except under special angles of incidence which lead to 'selective adsorption' in the potential well which is a few meV deep – not trapped at the surface and do not result in contamination of or change in the initial surface. The coupling of the incident atoms to the phonons occurs via the repulsive part of the interaction potential.

HAS is also a valid method for the investigation of the concentration of defects at a surface (steps, edges, dislocations, kinks). Statistically distributed defects result in diffuse scattering, in contrast to scattering from the periodically ordered lattice atoms. The cross sections σ for such processes are of the order of the gas phase interaction cross sections between helium and the defect atoms (100 Å^2). Thus the method is nondestructive, but has a sensitivity that allows one to detect defect concentrations of less than 1% of a monolayer.

The statistical adsorption of atoms (Volmer–Weber growth, Figure 3.8) results in a decrease in the specular scattering from the ordered substrate surface as the adsorbates result, as statistically distributed defects, in an increase of diffuse scattering events.

The decrease in intensity of the specular reflection is shown in Figure 4.42 for coverage of the mica surface with an alkali layer. One observes with increasing coverage an increase of scattering intensity following the initial decrease. The further development of the growth curve depends in the case of alkali growth on mica, strongly on surface temperature. At low temperatures (Figure 4.42a) the thick alkali layer grows in a

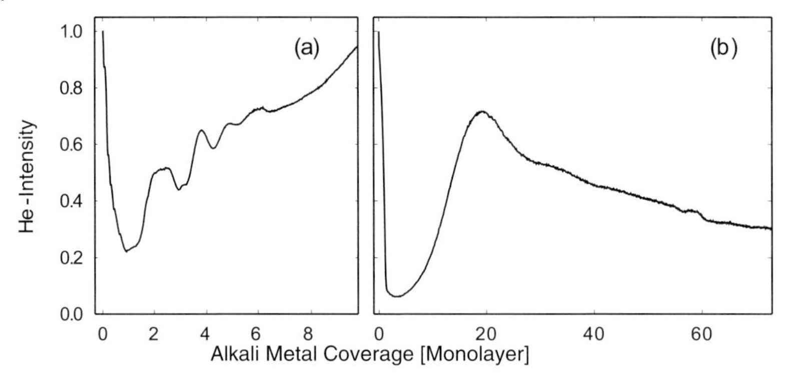

Figure 4.42 Specular HAS intensity as a function of coverage of an insulator surface with an alkali metal. (a) Surface temperature 53 K, $k_i = 6.4\,\text{Å}^{-1}$; (b) 150 K, $k_i = 7.5\,\text{Å}^{-1}$.

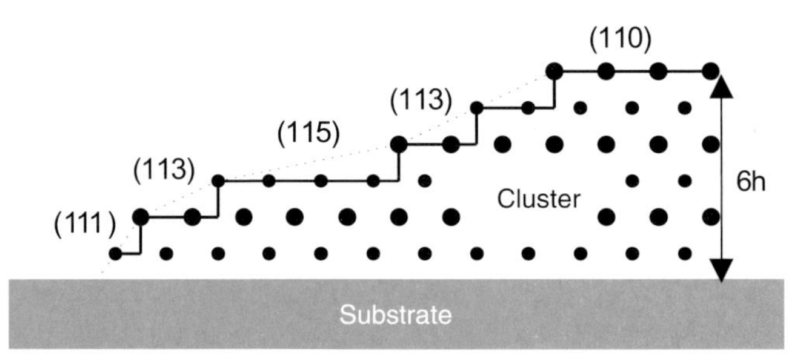

Figure 4.43 Microscopical view of a large alkali cluster on a substrate. Vicinal faces are shown, which contribute to the broading of diffraction patterns. The atom step height is $h = 0.21$ nm.

statistical layer-for-layer growth mode since the mobility of the atoms on the surface is restricted. Characteristic oscillations are observed: maximum scattering intensity is obtained at completion of a monolayer alkali atoms. For higher temperatures (Figure 4.42b) the scattered intensity increases since the reflectivity of the metal surface is much higher compared to that of the mica surface, but no oscillations (and thus no completed, 'perfect' monolayers) are observed.

Angular distributions show that the diffraction spectrum is reduced to a specular reflection peak which is strongly broadened. The broadening results from the growth of alkali clusters on the surface. At the tilted (vicinal) facettes of these clusters scattering is observed. Figure 4.43 shows schematically the microscopic structure of such a single crystalline cluster. A typical force- microscopy image is shown in Figure 6.30b.

4.5 Emission Methods

Information on the chemical structure of the surface can be obtained by methods which result in the emission of electrons having characteristic energies after irradiation with high energetic photons or particles. In photoelectron spectroscopy with deep ultraviolet or X-ray photons (UPS or XPS, respectively) electrons of the atoms A of the solid are excited from the valence bands (UPS) or from the K- and L-shells (XPS) into the continuum,

$$h\nu + \mathrm{A} \rightarrow \mathrm{A}^{+} + \mathrm{e}^{-}. \tag{4.35}$$

The subsequently emitted electrons are energetically analyzed by a spectrometer, for example, a cylindrical mirror analyzer (CMA). In the most basic arrangement this analyzer consists of two cylinders surrounding the target and the electron gun. Emitted electrons from the target enter the region between the cylinders through a small aperture and are repelled by the outer cylinder to exit the inner cylinder through a second aperture and hit an electron multiplier, which is mounted in line with the target. The position where they hit the multiplier is given by their kinetic energy and the voltage between the two cylinders. The entrance and exit apertures of the inner cylinder determine the transmission and thus the energy resolution of the CMA. An energy resolution of better than half a percent is easily obtained with transmissivities of the order of 10%.

From the energetic position and the intensity of the photoelectron maxima the abundance of specific elements at the surface can be deduced. From the shift with respect to the photoelectron maxima of free elements one obtains information about the binding strengths and positions of the elements at the surface. The line shapes in UPS allow one to determine the population density of the energy bands, and the angular dependence of the UPS intensities provides information about the electronic band structure.

Information about the short-range order around the surface atoms of selected elements can also be obtained via irradiation with X-rays with energies above the absorption edge. If one measures by use of an energy analyzer, the emission rate of photoelectrons as a function of the photon energy in discrete, element-specific energy windows, then one observes oscillations in the case of short-range order (SEXAFS, 'surface extended X-ray absorption fine structure'). These are caused by scattering of the photoelectrons which are emitted from the initially excited surface atom, at neighboring atoms. Since this is a coherent scattering process, characteristic interference phenomena appear. The same kind of interferences are observed as oscillations in the emission cross section as a function of

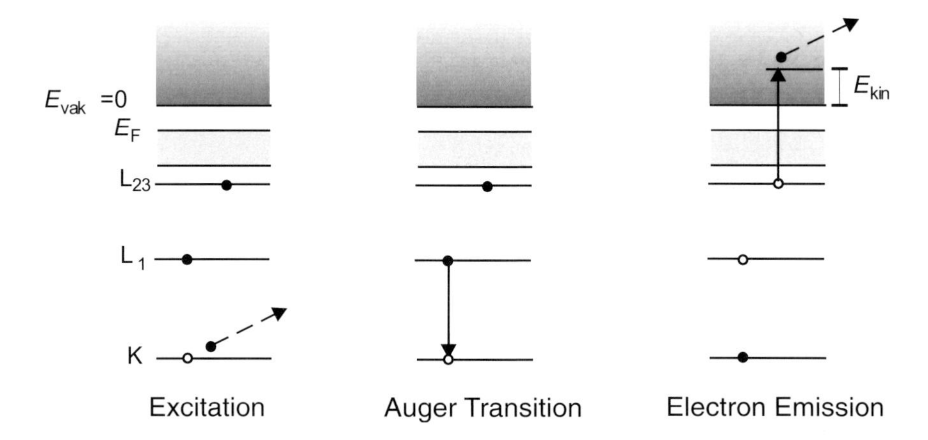

Figure 4.44 Term scheme for an Auger spectrometer. E_F is the Fermi energy, that is, the energy up to which the valence band is filled with electrons, E_{vak} is the vacuum energy of the free electron and E_{kin} the kinetic energy of the electrons following Auger ionisation.

electron and thus photon energy. From the amplitude of the oscillations, average atom distances and from their spacing the average number of interacting neighboring atoms can be deduced. Hence the method serves mainly for structural investigations. Close to the absorption edge (NEXAFS, 'near edge EXAFS'), one might obtain additional information about the electronic properties of the atoms, since the emitted photons can subsequently induce additional resonant excitations.

A sensitive chemical analysis of surface adsorbates, which also provides a convenient manner of monitoring surface cleanness and calibrating surface coverage, is made possible by Auger electron spectroscopy (AES). As a by-product one obtains quantitative information about the cleanliness of surfaces and/or the degree of coverage with nanoscaled aggregates. The ionization of the inner electronic states (e.g. of the K-shell) of the surface atom via irradiation with high-energy electrons (several keV) results in a hole in the K-shell, which is filled by the transition of an electron from, for example, the L_1-shell (Figure 4.44). The difference in energy is used to release an electron from a higher shell (L_{23}) from the solid. The kinetic energies of these Auger electrons, which are determined again with a CMA, are given by the differences between the relevant atom binding energies E_b:

$$E_{kin}(KL_1L_{23}) = E_b(K) - E_b(L_1) - E_b^{eff}(L_{23}), \tag{4.36}$$

where the superscript "eff" means that the ionization potential takes into account that the atom has already been ionized by the loss of an inner-shell electron. Since the kinetic energies depend only on the energy differences between electronic states in the atom, they are element-specific. Moreover, their absolute values are between 160.000×10^{-19} J (100 eV) and 1600.000×10^{-19} J (1000 eV) and hence their penetration depth is small (0.5–1 nm) and the method is truly surface-sensitive.

The intensity of the Auger lines is given by

$$I_{KL_1L_{23}} = I_0 \cdot \sigma_K(E_0) \cdot P_{KL_1L_{23}} \cdot T(E_{\text{kin}}) \cdot D(E_{\text{kin}}), \qquad (4.37)$$

with $\sigma_K(E_0)$ the ionization cross section of the K-shell, $T(E_{\text{kin}})$ the transmission function of the CMA, $D(E_{\text{kin}})$ the detection sensitivity of the electron multiplier and $P_{KL_1L_{23}}$ the Auger decay probability. Especially for elements with larger masses, radiative relaxation (i.e. fluorescence) competes with the Auger yield and makes an analysis of absolute intensities difficult. The measured Auger spectrum is thus generated by elastically scattered electrons (typical excitation energy 1.600×10^{-16} J (1 keV)) and a broad distribution of inelastically scattered electrons which have lost their energy into elementary excitations of the solid (Figure 4.45a). Therefore one usually modulates the outer cylinder of the CMA with a small voltage and detects phase-sensitively the first derivative $dN(E)/E$ of the Auger electron signal. Groups of sharp peaks at predefined energies then provide a 'fingerprint' of the elements forming the surface under investigation.

In addition to a chemical analysis of the surface, AES also enables one to obtain a depth profile analysis of the chemical composition of the solid. A possible approach is to change the angle of incidence of the primary electrons on the surface and thus to modify the effective penetration depth. In order to obtain information about deeper layers one has to remove the surface via ion sputtering before performing the AES measurement. This method is of special interest for nanoscaled layered systems as shown in Figure 4.46. Figure 4.46 is an elemental depth analysis of a metal/oxide/semiconductor system, obtained by subsequent sputtering of 8 nm thick layers.

The 75 nm thick silver layer on the surface is visible, which transforms into a nominally 25 nm thick copper layer. Apparently the copper has penetrated the oxidized silicon layer during deposition. Such detailed depth information allows one to optimize growth processes of ultrathin layers and thus to improve the definition of borders between layers of different materials. This is of tremendous importance for the use of ultrathin layers in nanoscaled electronic elements.

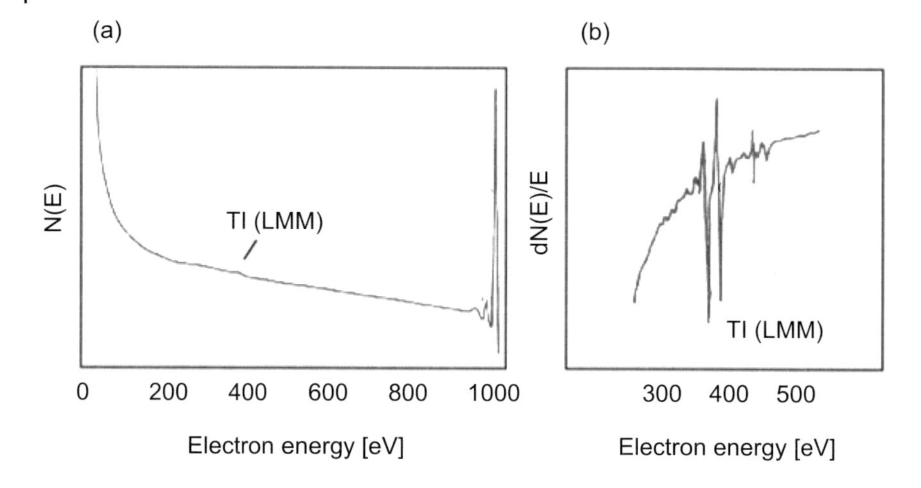

Figure 4.45 (a) Typical AES spectrum for the titanium LMM line, detected with a CMA. The primary electron energy is 1.600×10^{-16} J (1 keV). On the right-hand side a maximum of elastically scattered electrons is observed; plasmon excitations are seen at low energy loss. (b) Dericated of part of the spectrum in (a).

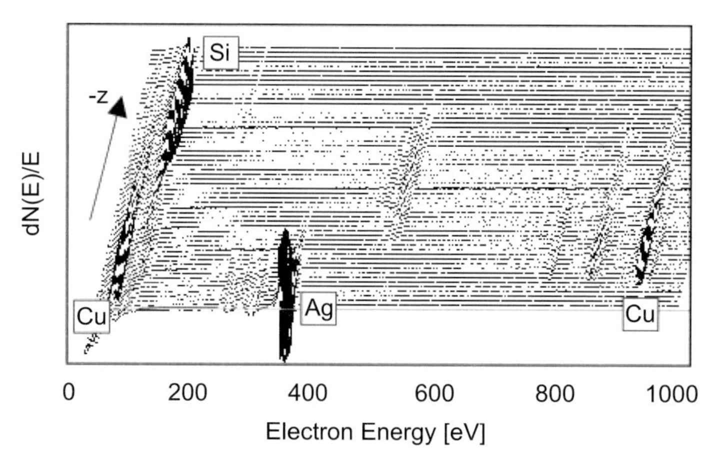

Figure 4.46 Depth profile analysis via AES. Auger spectra for different depths $-z$ are shown in a 75 nm Ag/25 nm Cu layered system on a silicon substrate, which is covered with natural oxide. The average depth decreases by $\Delta z \approx$ 8 nm between subsequent steps [322].

Problems

Problem 4.1 In order to obtain STM images with atomic resolution, an atomically sharp tip is needed. Why can a tip that has touched the surface still provide atomic resolution? How much does the tunnel current increase, if the gap between the tip and the sample is decreased by 1 Å? Assume a barrier height of 8.000×10^{-19} J (5 eV) and a small electron energy ($\approx 0.000 \times 10^{-19}$ J (0 eV)).

Problem 4.2 STM relies on tunneling of electrons and thus to first order works only for conducting surfaces. Why, despite this, can one image non-conducting surfaces and which limits apply?

Problem 4.3 Calculate the spring constant of a typical contact mode AFM cantilever. The deflection of the cantilever end-point $u(L)$ upon exerting a force F is given by:

$$u(L) = \frac{L^3}{3EI}F \tag{4.38}$$

where E is Young's modulus and I is the plane moment of inertia. For a rectangular cross-section, $I = (wh^3)/12$. The material is silicon (E= 150 GPa), L= 450 μm, w= 40 μm, and h= 2 μm.

How does this compare to the spring constant of a typical macroscopic spring?

Problem 4.4 Why is STM in principle a technique with much better lateral resolution compared to AFM? Why are AFM images much easier to interpret compared to STM images?

Problem 4.5 What kind of material properties can be deduced from XPS measurements?

Problem 4.6 What limits the application of electron microscopy to the study of insulating substrates? How can one overcome that problem?

Problem 4.7 Discuss why multiphoton excitation in imaging microscopy to first order leads to a decrease in resolution but in practice increases resolution.

Problem 4.8 Assume a p-polarized light beam traversing from a glass slab (refractive index n= 1.4) into air. Which angle is necessary for the glass slab surface in order to avoid reflection losses at the interface? What main assumption regarding the interaction between light and matter is behind the existence of the Brewster angle?

Problem 4.9 Assume a zone plate for a wavelength of 1 nm with a focal length of 1 mm. The lithographically achieved zone width is 20 nm. How many rings does the zone plate have and what is the radius?

5
Nano Architecture

In addition to the mainly surface-based growth methods for nanostructures that have been discussed in Chapter 3, there is a variety of methods available to generate three-dimensional nanostructures. These are briefly discussed in the present chapter; however, a lot of structures such as nanocomposites [323] have to be omitted.

Important processes to achieve a nano architecture are:

- The LEGO principle: simple basic elements, elementary rules and a simple building principle. Following the LEGO principle at the atom level nowadays even single atoms can be moved and arranged to new structures using the scanning- tunneling microscope. A well-known example are quantum corrals made from metal atoms (Figure 3.7), but also rows of atoms, which might be used as a 'nano-Abakus' [324]. The elementary rules are not at all trivial, but well known from quantum mechanics. In many cases the overall arrangement is just the sum of the individual parts. The LEGO principle is a physics based method of assembly from simple units, which works also with units larger than atoms. In addition to scanning methods [325] a variety of other, more indirect methods such as electrophoresis can be used for assembly [326].

- Cluster formation: The self organized growth of atomic or molecular units leads to nanoscaled structures with special structural and electronic properties. For example, aerosols from sulfuric acid, water and organic materials which are generated in the atmosphere and have diameters of a few nanometers, are condensation points for micrometer sized atmospheric cloud droplets.

  Since long clusters have been investigated with a variety of physical and chemical methods. This has led to very interesting (and sometimes surprising) observations such as the discovery of the extremely stabile C_{60} cluster (Section 6.5) or light emission from silicon clusters, a phenomenon that is closely related to porous sil-

Basics of Nanotechnology: 3rd Edition. Horst-Günter Rubahn

ISBN: 978-3-527-40800-9

icon. A general overview over atomic and molecular clusters can be found in [327,328].

In addition to their intrinsic properties clusters are also of interest as cages in which reactions run with well defined distributions of degrees of freedom. These cages can also be utilized for spectroscopic investigations at extremely low temperatures [329]. Especially large, superfluid ^{4}He clusters, generated via expansion of helium gas in a vacuum apparatus similar to that discussed in Section 4.4 have large application potential [330]. The temperature inside the cluster is 0.38 K, so that they resemble ultracold nano refridgerators, which allows high resolution spectroscopy of large organic molecules [331]. These clusters are nanoscaled non-biological relatives of the micro- or nano-droplets that are used as 'magic bullets' for controlled application of medicine.

The next step in nano architecture with clusters is the combination of individual aggregates using conducting organic molecules. This can be achieved by formation of a coating shell around the metal clusters using self organizing molecules (cf. Figure 6.32). If the clusters behave as metallic quantum dots then from an electronic point of view, a periodic super lattice with extraordinary electronic properties can be generated, the unit cell of which consist of a single cluster plus the organic molecular shell [332]. The use of an appropriate solvant results in three-dimensional lattice structures with,for example, periodically changing metallic and semiconducting clusters (quantum dots) [333].

- 'Molecular Engineering', molecular nanotechnology and supramolecular chemistry: the fundamental properties of the final structures are already programmed on the molecular level. This can be achieved via self assembly on specific template surfaces [8] in the gas or the liquid phase. Supramolecular chemistry [334] studies the coaction of molecules (or molecular units) with tailored properties that allow them to build up complex structures in a self organized fashion. Appropriate basic elements for the architecture of supramolecular assemblies, 'giant molecules' or chemical nanostructures are rodlike molecules [335] or oligomers and polymers with well defined length and construction [336, 337]. From those units spheres, double spirals or lattice structures form independently. The chemistry of membranes, enzymes and proteins, bio-organics and bio-anorganics, can also be assigned to the field of supramolecular chemistry. Here, reactions occur in a medium determined by water, polar surfaces and less polar surfaces (e.g. the membranes).

A more recent example of supramolecular selforganization are octohedral metal complexes with octupolar symmetry, which are of interest for nonlinear optical applications [338]. These self fabricating molecular units are packed subsequently via 'crystalline engineering' with optimization of the distances between the units in the respective crystal planes.

A further promising construction plan is the generation of dendrimers [339], which possess multifunctional properties owing to the large multitude of branches and their fractal structure. With the help of dendrimers a liquid soluble, macromolecular architecture can be generated in which the individual units remain in the nanometer range (10–20 nm). A direct comparison of dendrimers with conventional linear polymers has shown that such architecture results in improved properties – for example, improved nonlinear optical activity [340]. The polymers have been produced from the same units as the dendrimers.

- Selforganized chemical 'supra'-interactions: This class includes the fabrication of crystals from single wall carbon nanotubes (SWNT, cf. Section 6.5) [341]. This might be achieved by a mixture of C_{60} and Ni (as catalyst) transferred through 300 nm diameter holes. This mask has to be positioned laterally with a precision of 1 nm, meaning that the material mixture and thus also the subsequently growing nanotubes can be placed very precisely on well defined binding sites of the underlying substrate surface. The growth process consists of evaporation of layers of C_{60} and Ni with subsequent heating within a magnetic field. The C_{60} molecules combine to perfect aligned nanotubes in the form of a nanotube crystal. The mask and an appropriate substrate avoid lateral diffusion during the heating phase, while the magnetic field results in a good alignment of the tubes.

5.1 Layered Systems

Three-dimensional nanoscaled structures can also be generated by adding two-dimensional nanoscaled layers of different materials on top of each other. The higher the degree of structure formation, that is, the thinner the layers, the higher the demand on the microscopically well-defined (epitaxial) growth of the layers. It is usually found that the optical and electronic properties of the new, nanoscaled materials depend to a large extent on the microscopic order inside and between the individual layers.

Layered systems of importance for micro- and nanoelectronics are usually based on semiconductor materials (e.g. silicon), on which further semiconductors or metals and insulating intermediate layers are evaporated. With the help of lithographical techniques lateral structures are etched onto these layered systems. While the minimum lateral structure size has been improved down to a few ten nanometers by using shorter wavelengths and advanced illumination technology, further improvement of the vertical structure formation is hindered by the intrinsic roughness and the defect density of the evaporated layers. Thus, there is a strong need for improved (i.e. smoother and defectless) insulator layers.

Especially oxidation and passivation of the silicon surface is an important part of the layered system in modern semiconductor technology [342], and especially here the basic steps of molecular adsorption, dissociation and reaction are less well understood [343]. The question arises whether it is possible to manufacture an appropriately thin, continuous silicon oxide layer to fulfill the dimensional requirements of nanotechnology.

A new approach for the generation of ultrathin oxide layers on silicon single crystals (single monolayer thickness) is to apply dissociative adsorption of oxygen at room temperature, followed by a short heating cycle [344]. Growth mechanisms for the oxide are discussed in [345]. The MOS (metal-oxide-semiconductor) properties of the ultrathin silicon oxide layers have been determined by evaporation of a thin silver film and synchrotron induced photoemission measurements [346]. Especially the tunnel resistance of the layers and the defect density of states are important parameters which define the usefulness of the method within micro- and nanoelectronics.

New perspectives for photovoltaic applications can be expected from the combination of polymers with C_{60} molecules ('molecular plastic solar cells' [347]). In these photo cells the photoelectric effect is not generated in expensive silicon elements [348], but with the help of semiconducting polymers. The method bases in principle on solar cells which have been sensitized by the incorportation of dyes [349, 350]. Very efficient in this context is the combination of a conjugated polymer[1] with carbon nanoclusters (especially C_{60}, see Section 6.5). Advantages of organic solar cells are, in addition to their potentially low prices, the very low

1) The photoactive polymer in Figure 5.1 is methoxy-dimethyloctyloxy-poly-p-phenylen-vinylen; MDMO-PPV.

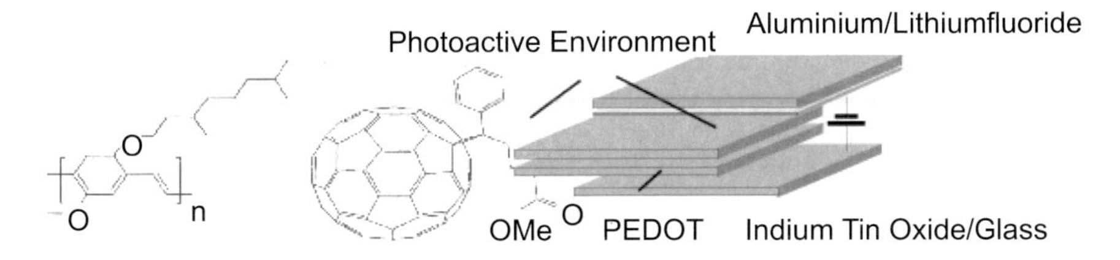

Figure 5.1 Schematic of a 'molecular plastic solar cell'. An 80 nm thick aluminum layer is evaporated as upper electrode onto a 0.6 nm thick Lithium fluoride layer. Between the lower ITO (indium tin oxide) electrode and the active layer (thickness 100 nm) is a 80 nm thick polymer layer. The chemical composition of the active layer, including C_{60} molecules, is indicated on the left-hand side. Reprinted with permission from [347]. Copyright 2001 American Institute of Physics.

weight, chemical inertness and mechanical flexibility. Disadvantageous so far has been the relatively small conversion efficiency (at maximum 6%) for solar light – about a factor ten below that of the best available single crystalline silicon solar cells. Single crystalline silicon solar cells result in efficiencies up to 24%. For large scale applications about 15% are achieved.

An optimization of the microscopic structure of the involved components results in a significant increase in the efficiency of the molecular plastic solar cells. Figure 5.1 shows the set up of a plastic solar cell, which exhibits a total conversion efficiency of $\eta_{AM1.5} = 2.5\%$. The conversion efficiency is defined as

$$\eta = \frac{P_{electric}}{P_{light}} = \frac{V_m I_m}{P_{light}} = \frac{V_{oc} I_{sc} FF}{P_{light}} \quad . \tag{5.1}$$

Here, $P_{electric}$ is the electrical power at optimum power use. Optimum power use results from power use at fitted working resistance (maximum voltage V_m and current I_m) modulo filling factor FF, which describes how much zero voltage V_{oc} and short cut current I_{sc} deviate from the optimum value. For the input power P_{light} one usually applies 'AM1.5': the intensity distribution of the laboratory light source corresponds to solar light which is irradiated through an air mass that is 1.5 times larger than that for normal incidence (the sun is at 41.8 degrees above the horizon).

The relatively low efficiency of the plast solar cell is mainly due to the solar like spectrum and the weak spectral overlap with the absorption bands in the organic materials. For specific wavelengths the efficiency can reach up to 100%.

In general the efficiency of systems with organic compounds depends very much on the molecular morphology, since the morphology influences to a large extent mobility of charge carriers and orientation of tran-

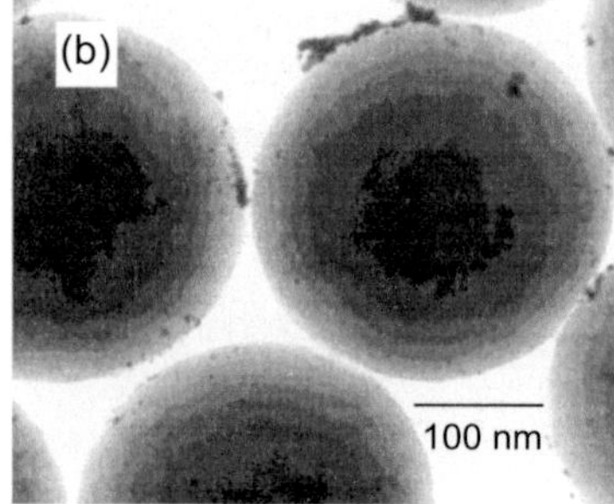

Figure 5.2 Macroporous materials from colloidal templates. (a) SEM image of a three-dimensional template from colloidal glass spheres. (b) TEM image of individual glass spheres with gold nanocrystals, which have been attached chemically to the glass spheres via thiole groups. Reprinted with permission from [351]. Copyright 2000 John Wiley and Sons.

sition dipoles. Special care must be taken to i) avoid the phase segregation of the C_{60} doped molecules in the clusters; ii) generate smooth interfaces to the electrodes; iii) maximize the interaction between the conjugated polymer strands. These requirements can be satisfactorily fulfilled by addition of an appropriate solvent and deposition via spin casting[2][347].

5.2
Colloidal Solutions and Crystals

Colloidal solutions, including gold and silver particles with diameters of a few nanometers, were used by Roman artists [328] to give glasses different colors in transmission as compared to in reflection. The brilliance of medieval church glasses is also due to incorporated nano particles with very high scattering cross section for incoming light. More detailed investigations have shown that the color of the solutions (in reflection) varies from red (25 nm) to green (50 nm) to yellow (100 nm) with increasing Au particle size, while it changes from blue to yellow for silver particles (40–100 nm). The aggregation of the particles in the solution also influences the effective color: for silver particles the color changes from yellow to red with increasing degree of aggregation.

Sedimentation, filtration or even simple heating of colloidal solutions results under appropriate conditions in a self organization process in

2) 'Spin casting' or 'spin coating' denotes the spraying onto a rotating substrate.

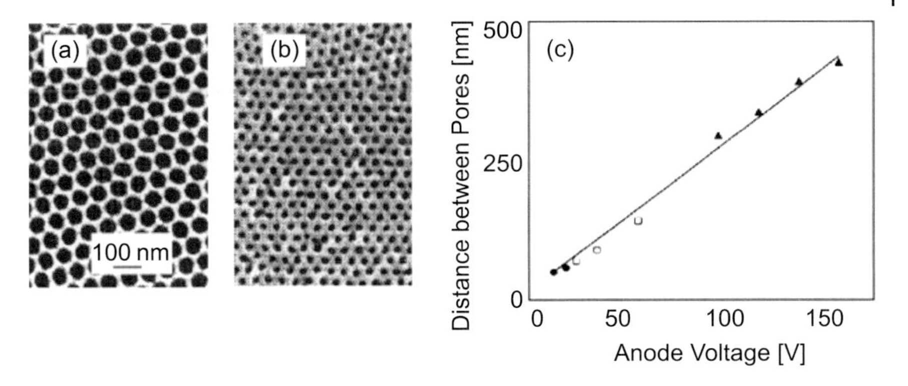

Figure 5.3 Hexagonal arrays of pores in aluminum. (a) Average pore distance 95 nm, (b) average pore distance 60 nm. (c) Linear dependence between pore distance and anode voltage. The symbols represent treatment with various acids during anodisation. Reprinted with permission from [353]. Copyright 1998 American Institute of Physics.

three-dimensionally ordered, densely packed arrays of glass or polymer spheres (Figure 5.2a). These arrays show order above several centimeters, possess a uniform thickness and after irradiation with visible light demonstrate characteristic diffraction phenomena [352]. In addition, these arrays are useful templates for the growth of macroporous metals (Figure 5.2b [351].)

As macroporous materials one usually defines materials with pore sizes between 200 and 1000 nm. Due to their large effective surface area (a 1 cm diameter, 0.5 mm thick macroporous gold film possesses a surface area of about 5 m^2) and the nanocrystalline structure of, for example, a macroporus metal, these materials can act as very effective catalysts. In order to generate a macroporous metal from the initial glass sphere template one first treats the spheres with a solution of mercapto silane, and so coats the surface of the glass spheres with sulfuric end groups (thiols). Following crystallisation on a substrate in the form of a colloidal crystal, gold nanocrystals (5 nm diameter) are bound to the thiols, again in solution. The strong, directed Au-thiol binding is used, for example, for the self organization of monomolecular organic thin films (cf. Section 3.2.2). By applying a voltage in an acid bath, these gold nanocrystals act as catalysts for the further growth of nanocrystalline gold. Finally, the initial glass spheres are removed with the help of hydrofluoridic acid. Obviously, the effective pore size is given by the radius of the glass spheres.

Materials with pore sizes below 50 nm are termed 'mesoporous'. Mesoporous materials serve as ultrafine sieves or, in nanophotonics, as photonic crystals.

5.3 Light Lattices

During interaction of light with matter, forces are induced that can lead to a change of particle movement. As early as 1619, Kepler recognized that radiation forces direct dust and ice particles of a comet tail away from the sun. The effect can be easily rationalized in the wave picture: the electric AC field $\vec{E}$ of light induces a force $e \cdot \vec{E}$ onto the electrons of the material, which consequently move within the magnetic field of the electromagnetic wave with velocity $\vec{v}$. Moving electrons in a magnetic field face the Lorentz force $-e \cdot \vec{v} \times \vec{B}$, which is directed perpendicularly to the electron movement. Hence, an average radiation pressure $\bar{P}$ is induced along the direction of the light, which equals the energy density of the electromagnetic wave. For a totally absorbing medium it follows:

$$\bar{P} = \frac{I}{c} \qquad [\mathrm{Nm}^{-2}] \tag{5.2}$$

where $I = (c\epsilon_0/2)E_0^2$ is the intensity of the wave with electric field[3] amplitude E_0.

In the photon picture a momentum can be associated with each photon

$$p = \frac{h}{\lambda} = \hbar k \quad . \tag{5.3}$$

The total radiation pressure equals integration of all photons, which irradiate a given area.

In the case of thermal light and macroscopic bodies the radiation pressure is very low. Let us, for example, try to move an absorbing sphere of 12 μm diameter and a mass of 10^{-12} kg with the help of focussed sun light. The solar constant[4] is of the order of 137 mWcm^{-2}, resulting in a radiation pressure of 4.5 μNm^{-2}. From that an acceleration of the sphere follows of 0.01 percent of the value of the acceleration of gravity.

For laser light including very high photon densities and/or for objects with nanoscaled dimensions (for example the particles that make up a comet tail) radiation pressure can play an important role. The absorbing

3) In the case of a totally reflecting medium the radiation pressure is twice that value.

4) The irradiance via the sun is not constant. From 136.3 mWcm^{-2} in 1600 it has been grown to 136.7 mWcm^{-2} nowadays.

sphere of the above example is now irradiated by a laser beam of 19 mW, focussed onto 6 μm[5]. The radiation pressure is then $1\,\mathrm{Nm}^2$, the sphere is accelerated with 13 g.

Besides light-induced movement of the sphere it should also be possible to trap micrometer-sized dielectric objects via the light forces that accompany focussing of laser beams. These trapped particles should be movable in aqueous solutions or against gravity [354]. These light tweezers play an increasingly important role in physics and biophysics research. For example, in addition to dielectric spheres [355] and gold clusters [356] also viruses and bacteria [357], living cells [358] and DNA-strands [359] have been trapped and manipulated. Recent developments include the levitation of microscopic objects against gravitational force via focussed light from optical fibers [360], the actuation of micrometer scaled 'light windmills' [77] (Figure 6.45) or the trapping of particle arrays via near field enhancement along nano-shaped metallic cluster arrays [361].

The trapping of dielectric particles close to the laser focus is a consequence of radiation pressure due to reflected and diffracted light beams. In Figure 5.4 the forces on a dielectric sphere close to a laser focus are plotted for a partial wave above the center of the sphere.

Part of the laser light is reflected from the sphere and results in a force F_r which is directed towards the center of the sphere with a component in z direction and a component in y direction away from the middle of the laser beam. The residual is diffracted into the sphere (F_b) and results in a force component in the direction of the laser beam center and a component in z direction. Upon exiting, the diffracted beam is again split into a reflected and a diffracted part. The force components for the exiting diffracted beam equal the components of the incoming beam. The beam which is reflected inside the sphere, however, has in addition to the force component in z direction a component, which is directed away from the laser beam center: this component compensates the force component of the incoming-reflected beam which is directed towards the beam center. Summing up, one ends with a net force in z direction towards maximum intensity of the laser beam.

For a partial beam below the center of the sphere the summation results in a force component in z direction, albeit directed away from the intensity maximum. Hence, we find a net force along the laser beam for a sphere in the beam focus. Outside of the beam center the sphere will be attracted towards the beam focus. If the sphere is immersed in a viscous medium, the maximum induced acceleration by the laser radiation pres-

5) The diffraction limited minimum focus area is approximately equal to the square of the laser wavelength.

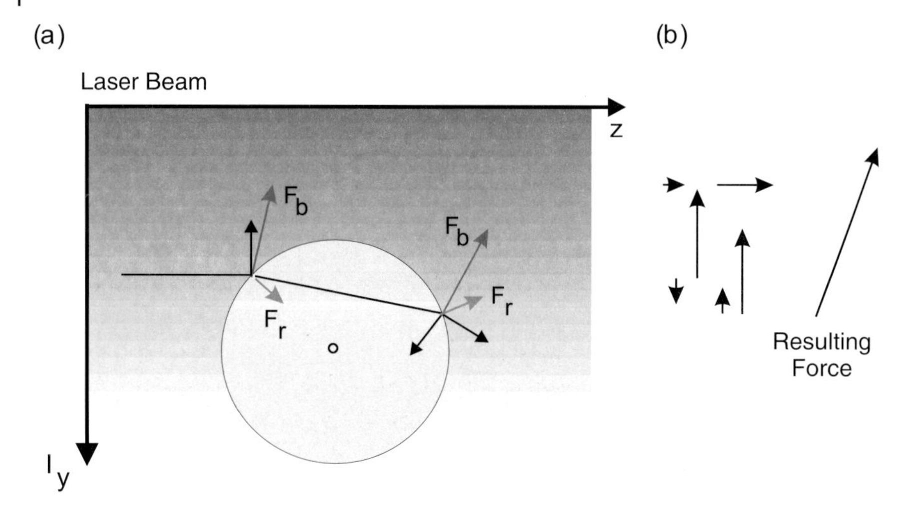

Figure 5.4 (a) Laser induced forces on a dielectric sphere with index of refraction larger than that of the environment of the sphere [354]. The sphere is situated in the exponentially decaying electric field of a laser beam, which propagates from the left to the right-hand side. (b) Resulting forces on the sphere.

sure is limited by the Stokes' friction force. The final speed of a sphere of radius $R < w_0$, which is illuminated by a laser of power P_L and beam waist w_0 in a medium of viscosity η is given by [354]:

$$v_f = \frac{0.12 P_L R}{3\pi c \eta w_0^2} \quad . \tag{5.4}$$

For water ($\eta = 10^{-3}\,\mathrm{Nsm}^{-2}$ at room temperature) and a $1\,\mu\mathrm{m}$ diameter sphere irradiated by a 1 W Laser, focussed onto $10\,\mu\mathrm{m}$, a velocity of one millimeter per second follows. In Figure 6.45 a micro machine is sketched, which is actuated by such light forces.

It is noted that objects in a gaseous or liquid environment always experience photophoresis [362], that is a non-uniform heating of the object, which results in a net momentum away from the beam maximum. The amount of heating depends on the absorptivity of the object, meaning that the complex index of refraction determines which effect dominates. Low absorptivity is crucial for simple trapping experiments.

On the microscopic level, the trapping of atoms by laser light can be understood via induced dipole forces and the isotropic character of fluorescence emission. The beam focus of a laser is an area with inhomogeneous field strength distribution, such that the force onto an atom with polarizability α is

$$\vec{F}_D = -\vec{p} \cdot \nabla \vec{E} \tag{5.5}$$

where $\vec{p} = \alpha \cdot \vec{E}$ is the induced dipole moment of the atom and $\nabla \vec{E}$ the gradient or the spatial variation of the electric field. The force averaged over an optical cycle is then

$$< F_D > = -\frac{1}{2} \alpha \nabla (E^2) \quad . \tag{5.6}$$

After some transformations [284] one finds as a function of the gradient of the laser intensity $I = \epsilon_0 c E^2$ taking into account the frequency dependence of the polarizability:

$$< F_D > \propto -\Delta\omega \nabla I \quad . \tag{5.7}$$

If the atom moves with the velocity $\vec{v}$, it is $\Delta\omega = \omega - (\omega_0 + \vec{k}\vec{v})$, where ω is the frequency of the laser field and ω_0 the resonance frequency of the atom. The spatial intensity distribution within the laser focus is usually Gaussian:

$$I(r) = I_0 exp(-\frac{2r^2}{w^2}) \quad , \tag{5.8}$$

with w the beam waist. That means that $\nabla I \propto -\vec{r} I(r)$ is directed radially outward. For negative (red) detuning of the laser with respect to the resonance frequency one obtains a radial force in direction $r = 0$: just as in the above discussed macroscopic case the laser beam axis represents a potential minimum, which attracts the atoms.

The depth of this minimum is shallow: for an intensity of $10^9\,\mathrm{Wm}^{-2}$ a sodium atom experiences a trap with a minimum well of $0.5\,\mu\mathrm{eV}$. If the atoms are in thermal equilibrium at a temperature of only 5 mK, they can escape that trap due to thermal motion. If one wants to trap atoms in such a potential minima with the intention of generating an artificial lattice structure from such trapped atoms, one has first to ensure that the temperature of the atoms is below 5 mK. A well know approach to this end uses optical absorption-emission cycles and 'Doppler-cooling'. see also Section 6.3. The cooling mechanism bases on the anisotropic exchange of photon momentum quanta. Hence the minimum achievable temperature is given by 'statistical heating' and is of the order of 0.25 mK for sodium (Equation 6.20).

If one superimposes two polarized light waves with fixed phase relationship within an ensemble of cooled atoms, standing waves with areas of maximum and minimum electromagnetic field strength are generated.

Within the areas of maximum field strength dipole forces bind the atoms. Thus, in principle it should be possible to fabricate regular, crystall-like structures in an atom gas by superposition of such light waves [363]. Two-dimensional crystalisation of rubidium atoms has been achieved by superposition of two mutually perpendicularly polarized, standing light waves in a magneto-optic trap [364]. The temperature of the atoms in the trap is of the order of 4 μK; thus the optical potential well depth is sufficient. The scheme can be extended to three dimensions by use of a further, phase-controlled laser beam. The periodicity of the resulting crystal has been verified by Bragg diffraction of a probe beam [365].

A 'light-bound' crystal has similarities with an ultracold atom gas of very high phase space density, which allows, for example, for Bose–Einstein condensation. Bose–Einstein condensation [366–369] occurs, if the particle density is high enough and the temperature of the bosons low enough that all of them populate the same ground state. In the wave description, which associates a matter wave with de Broglie wavelength $\lambda_{dB} = h/mv$ to a particle of mass m and velocity v, a single, 'macroscopic' wave function appears instead of a superposition of identifiable particles with individual wavefunctions. The optical analog is the coherent electromagnetic wave that is generated by a laser. It is, therefore, not surprising that an 'atom laser' has been fabricated on that basis [370].

Up to now it has not been possible to realize Bose–Einstein condensation with the help of a light lattice since the phase space density is not high enough (about 90% of the lattice sites are unoccupied). Instead one usually realizes the necessary density for the phase transition via optical cooling in a magneto-optical trap, MOT, with subsequent evaporative cooling.

In a MOT the attractive forces of an inhomogeneous electric field (generated by the laser) are combined with the forces that are generated by an inhomogeneous magnetic field. Due to the Zeeman-shift (red shift) of the energetic levels a strong restoring force is generated for atoms that prefer to leave the trap – this force increases with increasing distance to the trap and thus compresses the atoms in the trap. By appropriate timing of the magnetic fields the high energetic atoms are allowed to leave the trap, while the low energetic atoms remain ('evaporative cooling', similar to cooling of a hot liquid). That way the temperature of the ensemble is lowered below the limiting temperature for Bose–Einstein condensation, while the density remains sufficiently high.

It is finally noted that light lattices are interesting model systems for both fundamental research (statistics in multi body systems) and for applications, for example, in the field of nano-lithography with ultracold atoms [371].

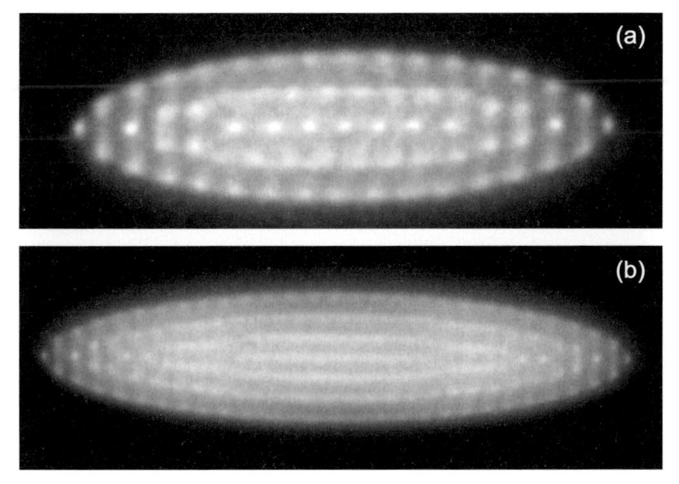

Figure 5.5 Coulomb crystals, produced from 200 (a) and 1000 (b) ions and visualized optically [373]. Reprinted with permission.

5.4 Coulomb Crystals

If one captures ions with an electric Paul trap[372], the system can stabilize under appropriate conditions into a crystal-like structure due to Coulomb interactions ('Coulomb crystal').

Molecular dynamics simulations have shown that the ratio of Coulomb energy between neighboring ions to average kinetic energy of the ions is the most influential factor. If this ratio is larger than 170 the ions are cooled enough to allow a phase transition into the solid state: the ions 'crystallize'.

In Figure 5.5 such Coulomb crystals with countable numbers of ions are shown. Ion densities of $10^8\,\text{cm}^{-3}$ and temperatures of a few milli-Kelvin inside the Paul trap are necessary for the generation of such structures.

With increasing particle density the Coulomb attraction also increases, meaning that the crystals can also exist at higher temperatures. It has been postulated that inside 'white dwarfs', that is, stars with high densities, at temperatures around 100 K, Coulomb crystals are formed. On earth this special kind of artificial crystals is a perfect playground for basic experiments on the interaction between low energy particles as well as on phase transitions in systems with countable numbers of particles.

Problems

Problem 5.1 Why are molecular nano architectures in general much more easily destroyed compared to the molecular building blocks they are made from? Name some types of non-covalent bindings.

Problem 5.2 What could one do to counterweight the effect named in Problem 5.1 and so make the molecular nano architectures more stable?

Problem 5.3 Nano architecture on the basis of light forces is limited by the diffraction limit. Show that the minimum focus area of a diffraction limited laser beam is approximately equal to the square of the laser wavelength.

6 Applications

6.1 Optics

6.1.1 Integrated Optics, Nanooptics and Nonlinear Optics

In order to find real world applications for integrated optics in the submicron size range one needs nanoscaled (i) light sources, (ii) detectors as well as (iii) nanoscaled switching and conversion elements and optical conductors. For nanophotonics these elements should ideally be combined on photonic integrated circuits (photonic chips), and so allowing for the possibility of combining them with further nanoscaled electronic or optic components.

In what follows the three basic components are visualized with the help of already existing elements. The rather widespread field of optical micro cavities [374], which has resulted in a series of microlaser concepts is excluded. Also the related cavity quantum electrodynamics [375] is not discussed. Basics and applications of optical phenomena which occur in the context of nanostructured materials (local field effects, photonic materials, optical nonlinearities, quantum wires and quantum dots) are discussed extensively in [376].

Light Sources

A maximum optical information transfer rate is achieved with coherent light sources. Hence 'nanolasers' are potentially very useful as primary nanophotonic components. Figure 6.1 shows a possible configuration. The 'nanolaser' possesses cross sections of the order of 100 nm. Commercial microlasers made from GaAs or GaN have cross sections of a

Basics of Nanotechnology: 3rd Edition. Horst-Günter Rubahn

ISBN: 978-3-527-40800-9

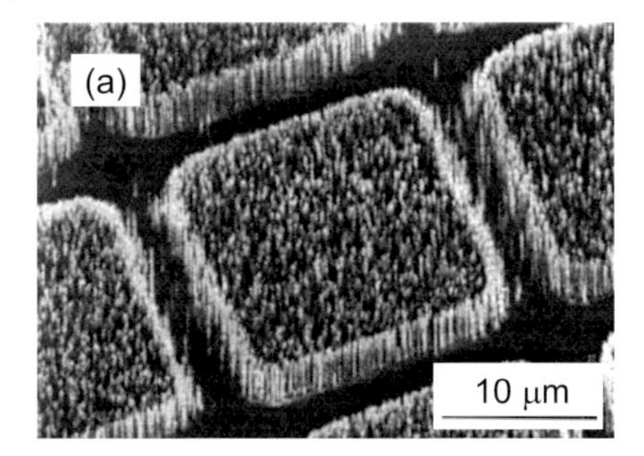

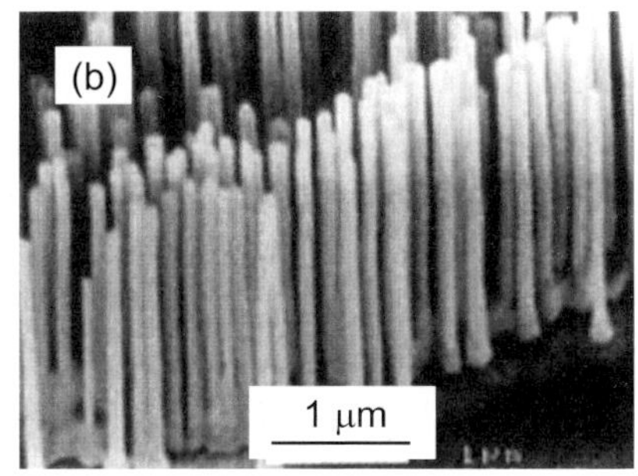

Figure 6.1 Nanolaser. a) SEM image of an epitaxially grown distribution of ZnO needles.b) SEM image with higher magnification. The individual needles act as nanoscaled lasers. Reprinted with permission from [378]. Copyright 2001 American Chemical Society.

few micrometers. The active gain material is tin oxide (ZnO), which has been grown epitaxially as crystallites on a sapphire substrate.

Tin oxide is a semiconductor with a wide bandgap (5.392×10^{-19} J (3.37 eV) at room temperature). After excitation with UV light possessing photon energies larger than the band gap it shows photoluminescence and also population inversion and laser activity if it exists as an ultrathin film or microstructure [377]. Crystal growth on sapphire can lead under specific conditions to the formation of long, upright oriented single crystalline needles or nanotubes (diameter 100 nm, length a few micrometer), which possess well defined hexagonal end faces (Figure 6.1). These end faces act as mirrors for photons, which are generated via pulsed laser excitation in the ZnO needles. As a result an intrinsic resonator is generated, which selects certain modes and allows a build-up of stimulated emission[1].

For the distribution of ZnO needles shown in Figure 6.1, scanning nearfield optical microscopy has only detected laser activity for a few percent of the needles. The reason for the large amount of non-lasing needles is that the reflectivity of the sapphire-surface oriented face of the needles is greatly influenced from the substrate. Even in the best case, the maximum number of resonator round trips is only 3, which has a negative effect on the mode spectrum and the divergence (spatial coher-

1) The building principle of this kind of laser is in general agreement with the building principle of the first optical laser, presented 1960 by T.Maiman: in that first laser a ruby crystal was polished and one side metallic coated, and so served as an active laser cavity.

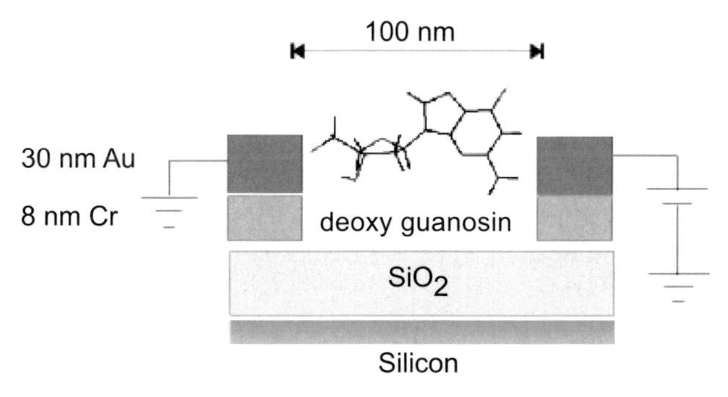

Figure 6.2 Biomolecular hybrid photodetector [380].

ence) of the emitted laser light. A potential application in nanophotonics would of course also ask for electrically stimulated laser action.

An alternative to the vertically oriented inorganic ZnO nanolasers are horizontally oriented organic nanolasers based on poly-phenylene molecules [176, 379]; see also Section 3.2.2. Such organic nanolasers also possess crystalline facets that allow resonator feedback, are made of a high gain medium and have a fortunate optical term scheme for stimulated emission. Due to their flat orientation on the surface these nanolasers, however, can be more easily contacted with electron and hole donating electrodes. Also, the lasing material can be 'fine-tuned' by modifying the molecular building blocks. Electrically pumped, tunable nanolasers might become a reality that way.

We finally note that a large variety of further micro-nanoscaled lasers exist such as self organized quantum dot lasers [17], which resemble the bottom-up variant of the quantum well diode laser.

Detectors

Sensitive, nanoscaled photo detectors can be generated with the help of biological materials. The DNA base deoxyguanosin, for example, is a semiconductor with a band gap of about 5.120×10^{-19} J (3.2 eV) if it occurs in the form of a self organized ultrathin film. In combination with lithographically structured nanoelectrodes a biomolecular hybrid-photodetector can be generated from this material (Figure 6.2).

To achieve this, one generates via electron beam lithography two Au/Cr electrodes with a mutual distance of about 100 nm on a SiO_2 substrate. The DNA molecules are dissolved in chloroform and dip coated onto the surface. During evaporation molecular wires are formed between the electrodes, which possess semiconducting properties. Irradiation with light results in emission of electrons. The detector has a sensitivity of about 1 AW^{-1}.

The huge advantage of such biomolecular hybrid detectors in comparison with the more bulky semiconductor products is that the photoactive material can be deposited onto the submicron structured surfaces with the help of a simple inkjet printer[2].

Doublers and Switches

Frequency Doublers Organic thin films are of huge interest for integrated nanooptic applications, in particular owing to of their potentially high nonlinear optical activity. The high nonlinearity results from the specific hydrocarbon based molecules with electron acceptor and electron donor end groups, which posses a strongly polarisable conjugated π-electron system. Dye-doped films have shown second order nonlinear optical polarizabilities of up to 10^{-13} esu [382]. If one integrates these films into ultrathin film waveguides one can fabricate light guiding frequency doublers with microscopic dimensions and high conversion efficiencies [117]. Examples for submicron scaled waveguides are discussed in Section 3.2.2. See also Figure 3.17.

In the regime of nano-sized elements or ultrathin films, the mutual interaction between the individual nanoscaled doubling elements can of course be problematic. In contrast to macroscopic frequency doublers, which are large compared to the wavelength of the frequency doubled light, the individual elements and also their mutual distances are in the present case small with respect to both the incoming and frequency doubled radiation. This results in interference phenomena which can lead to an enhancement of the frequency doubling probability, but also to destructive interference and thus a strong attenuation of the effect. The possible optimum mutual alignment of the elements towards a high macroscopic polarizability ('susceptibility') is in addition limited by electrostatic interactions. This has been observed, for example, for 'poled' polymer rods from nonlinear optical active material, resulting in a rather small macroscopic nonlinearity [383].

Recently developed crystalline organic nanofibers (Section 3.2.2) add new possibilities. It has, for example, been demonstrated that it is possible to grow well-oriented nanofibers on specific template substrates such as muscovite mica via organic molecular beam epitaxy. The nanofibers

2) Nanoscaled photodetectors on the basis of semiconductor quantum dots have also been developed [381]; their function is, however, optimum only at low temperatures (77 K, liquid nitrogen temperature).

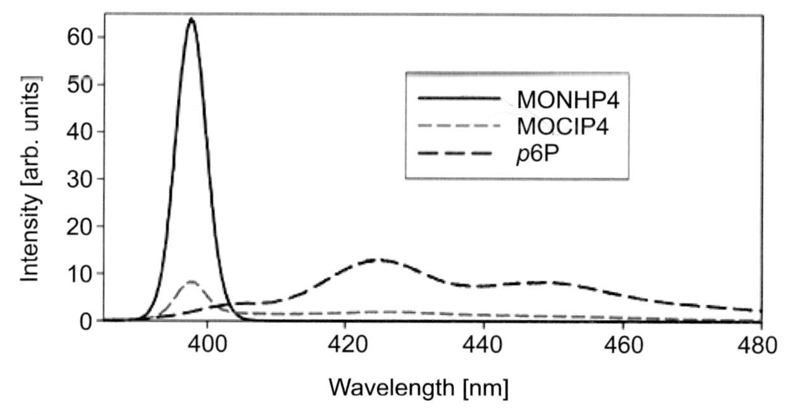

Figure 6.3 Optical emission spectrum of differently functionalized para-quaterphenylene molecule based nanofibers, induced by infrared femtosecond laser excitation. Asymmetrically functionalized nanofibers show pronounced second harmonic generation. J.Brewer and M.Schiek, NanoSYD, private communication, 2008.

consist of molecules oriented nearly parallel to the substrate surface and perpendicular to the long axis of the fibers with aggregate dimensions of a few hundred nanometers width, a few ten nanometers height and several hundred micrometers length. An important feature of this class of nanoaggregates in terms of frequency doubling is their crystalline perfection – due to their epitaxial growth process they resemble the optimum configuration for the given molecular constitutes, thus also demonstrate the optimum optoelectronic properties, just limited by the molecular building blocks themselves.

Such 'perfect' epitaxial growth imposes strong boundary conditions on the choice of substrate and adsorbate. Although para-hexaphenylene (p6P) molecules are very interesting in terms of photonic applications due to their high quantum efficiency, the hyperpolarizability is very low since they lack donor and acceptor groups. One way out of this dilemma bases on the asymmetrical chemical functionalization [173] of a para-quaterphenylene (p4P) block with electron push and pull groups, for example, methoxy and amino groups [175]. The p4P block is necessary to fulfill the epitaxial growth criteria.

Figure 6.3 depicts how the optical emission spectrum changes upon infrared excitation (800 nm) as one uses differently functionalized molecules. For p6P the spectrum consists solely of two-photon luminescence (TPL), while in the case of methoxy and amino functionalized p4P (MOP4NH2) it consists solely of second harmonic generated (SHG) light. If one uses chlorine instead of the amino groups one obtains both SHG and TPL.

Obviously, molecular tailoring provides one with high optoelectronic flexibility.

Such nanoscaled, integrated frequency doublers might allow one to use cheap, power saving and brilliant near-infrared laser sources such as easily integrable PBG (photonic band gap) VCSELs (vertical cavity surface emitting lasers) as the main source for intense, coherent blue light sources.

Switches In contrast to conventional semiconductor elements with resonant switching times of nano- or picoseconds one can realize with the help of organic thin films transient switches on a femtosecond time scale [384]. An example is a change in the index of refraction of the organic film, induced by a switching pulse. This change induces a phase modulation in a subsequent pulse. Alternativily a diffraction grating is written into an organic material, which switches the following pulse spatially between two subsequent wave guides (nonlinear optical directive coupler). The huge nonlinearity also allows one to use holographic diffraction phenomena ('Four Wave Mixing, see Section 4.3) on a time scale of femtoseconds. It should be possible to induce a phase conjugating mirror within an ultrathin film, which could be used for the correction of dispersion effects in optical data lines[385].

Tailoring of Optical Properties Periodic particle assemblies such as those discussed in Section 3.1.1 and Section 5.2 localize the near field and under certain conditions enhance the probabilities for Raman scattering [386] and optical frequency doubling. Thus the optical properties[3] of, for example, extinction spectra [387] or the surface plasmon enhanced near field might be tailored [388]. Especially the latter appproach is very interesting regarding possible applications in the field of nanooptics. Figure 6.4 shows a chain of gold clusters on an indium-tin oxide (ITO) surface, obtained with a photon scanning tunneling microscope (PSTM). The clusters, with typical dimensions of $(100 \times 100 \times 40)\,\mathrm{nm}^3$, were generated via electron beam lithography. The distance between individual clusters is about 100 nm, that is, much smaller than the wavelength of illuminating light (633.8 nm). The PSTM measures with the help of an optical fiber tip the emitted light intensity following object illumination within the evanescent wave. Thus it is sensitive to the optical near field.

Comparison with a numerical simulation (Figure 6.4c) reveals that the maximum light intensity does not appear at the position of the clusters

3) The size dependence of optical properties can be used to excite particles of certain size using laser light, which might lead to desorption of the particles; see [114].

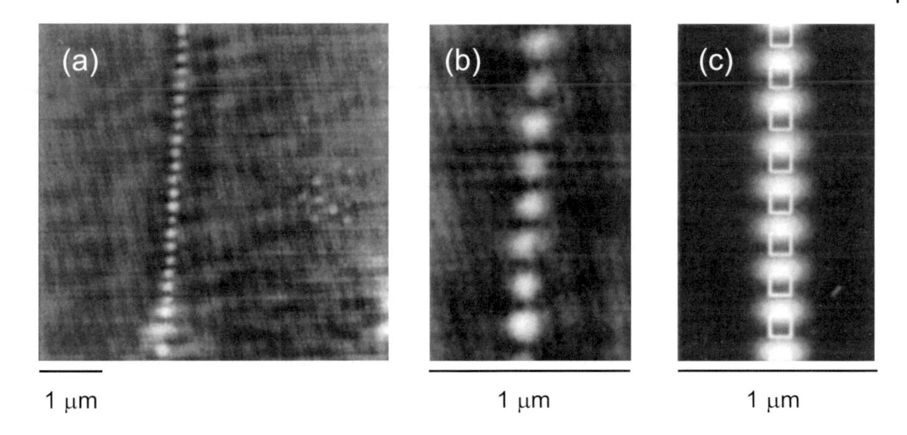

Figure 6.4 (a) Linear chain of gold nanoparticles on an ITO surface. (b) Magification of image (a); (c) numerical simulation. The position of the clusters is denoted by white squares. Reprinted with permission from [389]. Copyright 1999 The American Physical Society.

but between them. Due to the mutual interaction (plasmon coupling) an eigenmode of the complete chain of clusters is excited, which leads to a 'squeezing' of the optical field: the observed light spot is much smaller (85 nm diameter) than an individual cluster (250 nm). In fact it would not be possible to separate individual clusters in the chain without the squeezing effect.

6.1.2
Optical and Magnetic Data Storage

Most data storage still occurs using magnetic media with steadily increasing storage density, although RAMs (random access memories) do process on a semiconductor basis (DRAM, dynamic RAM). The generation of nanoscaled periodic magnetic structures (cf. Section 3.1.1) might allow in the future one to implement MRAMs (magnetic RAM) instead of DRAMs. This would implement an increased storage capacity, current less storage and very low energy consumption. In order to use MRAMs, the magnetic quantum dots in the periodic pattern have to be accessed individually, have to be read out and the magnetisation direction has to be changed. This is at present difficult, since the magnetic pattern possesses in general no rotation symmetry, whereas the high read and write speeds of present hard disc memories rely on a high rotation speed. However, in principle the generation of rotation symmetric, self organized patterns of magnetic quantum dots is feasible.

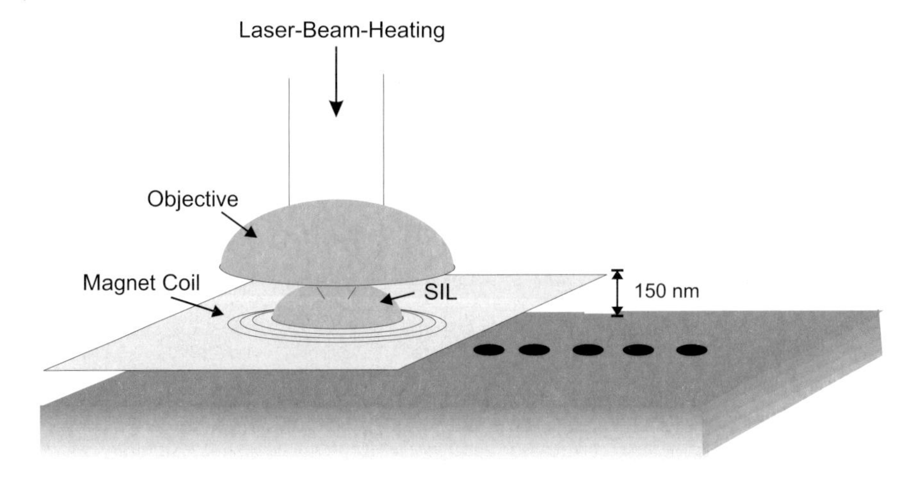

Figure 6.5 NFR data write head ('flying head').

We will not discuss in any further detail the many aspects of magnetic data storage and nanoscaled magnetic effects; rather, we direct the reader to an overview which can be found in [390]. We note, however, that the generation of ultrathin magnetic films and the corresponding new magnetic effects (the film thicknesses are of the order of the average mean free path of the conduction band electrons) have had a significant effect on present computer technology. One good example is the dramatic change of electric resistance as a function of external magnetic field in film systems made from a ferromagnetic (e.g. 2 nm Cobalt), a non-magnetic (e.g. 2 nm Copper) and a ferromagnetic layer (2 nm Cobalt), bound by an antiferromagnetic layer (nickel oxide) (GMR, 'giant magnetoresistance' [391,392], recent nobel price in physics). Most of the currently used reading heads for hard discs employ such GMR thin film sensors and achieve reading rates of about 10^8 Hz. Exchanging the conducting non-magnetic layer with an insulator layer might increase sensitivity and thus the possible read out rate (TMR, tunneling magnetoresistance).

Besides magnetic storage media, laser structured media (CDs and DVDs) are also widely spread. An interesting approach to increasing the surface data storage on such media is data recording in the near field (NFR, 'near field recording'). Figure 6.5 shows a corresponding data write head.

Reading occurs with the help of a laser beam that heats the surface a few hundred degrees Kelvin within a diffraction limited spot with a high heating rate of a few $10^{11}\,\mathrm{Ks}^{-1}$. Correspondingly, a pulsed magnetic field is sufficient to locally magnetize the surface. The magnetization coil is integrated into the writing head and thus the magnetization can be changed on a very short time scale ('direct overwrite'). The maximum data density is limited by the size of the diffraction limited laser focus diameter, which localizes the magnetization spot. This diameter depends on the numerical aperture (NA) of the imaging objective (cf. Equation 4.5).

The NA might be enlarged by the use of an immersion oil, that is, taking advantage of total reflection (Figure 4.4). In the NFR-data write head this is enabled by use of a lens made of a high index of refraction material n_{SIL} (SIL: 'solid immersion lens'), which is situated in the optical near field of the surface (depending on the applied laser wavelength approximately 150 nm). This short distance between data write head and medium is achieved by an air cushion. From Equation 4.5 it is seen that for $n_{SIL} = 2$ the spot diameter can be reduced by a factor of 2. An additional increase of data density can be achieved by the use of shorter wavelengths (e.g. a blue or green write laser) and special write patterns (e.g. overlapping patterns) [393].

6.1.3
Photonic Crystals

Semiconductors are especially useful materials for the electronic industry since the electronic band gap allows control of the transport of conduction electrons. The band gap between valence and conduction bands spans an energy range inside which the electrons cannot move through the solid. If the electron energy, however, is large enough to overcome the gap the electrons can move freely through the conduction band.

A similar effect can be obtained using light if one fabricates a 'photonic band gap' in a dielectric medium. By generating an appropriate three-dimensional periodic pattern of macroscopic regularly distributed scatter points ('artificial atoms' of a newly designed material) a band gap for incoming photons is fabricated (Figure 6.6). Hence, one or several frequency ranges $\nu \cdot a/c = f$ (a is the lattice constant), where $f < 1$, exist, inside which photons – independent of direction – cannot propagate. Incoming light is modulated in propagation and frequency spectrum. In general such materials are called 'photonic crystals' [394–397].

The validity of this approach can be seen, for example, on evolutionary optimized structures such as butterfly wings (Section 6.4). The wings consist of periodic arrays of crossed posts in face-centered cubic structure (fcc-structure) with typical dimensions in the sub-micrometer range. Light transmission and reflection are, thus, modified that way, and the

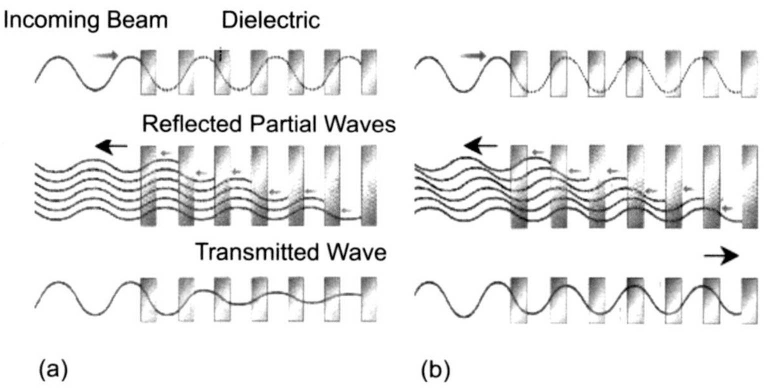

Figure 6.6 An one-dimensional photonic band gap [398]: (a) incoming light is reflected in-phase and annihilates the transmitted light. (b) Outside of the band gap the reflection experiences a phase shift, which means that the transmitted light is attenuated only weakly.

butterfly wing shows strong iridescence. The angular dependence of this effect is an indication of an incomplete photonic band gap. In order to obtain a complete photonic band gap, light has to be perfectly reflected for a certain range of wavelengths and independent of propagation direction in the crystal.

As seen from the example of the butterfly wing any material can act as photonic crystal, the index of refraction of which changes periodically on a typical scale of half a wavelength. However, for the generation of a complete band gap the requirements on the crystallographic structure of the artificially changed dielectric are non-trivial. It has been shown that the diamond structure (tretahedral) leads to the most promising photonic crystals. An example for such a three-dimensional periodic dielectric structure with a *complete* photonic band gap is shown in Figure 6.7 [399]. The structure consist of an array of periodic dielectric tubes in air plus air cylinders in a dielectric. These cylinders represent a face-centered cubic (fcc) in (111) direction oriented lattice of low index material, inserted into a high index material. The air cylinders have radii of 0.293 a (a is the fcc lattice constant) and heights of 0.93 a.

If a defect inside the crystal lattice exists, that is, if the symmetry is broken, then light can also propagate within the band gap of the photonic crystal. A point defect acts as a cavity at which the photon can be localized, a line defect as a waveguide and a planar defect as a perfect mirror. Since the periodic lattice has been fabricated completely artificially with respect to structure and material (e.g. by imprinting a nanoscopic hole pattern), the defects can be fully manipulated, resulting in a uniform and

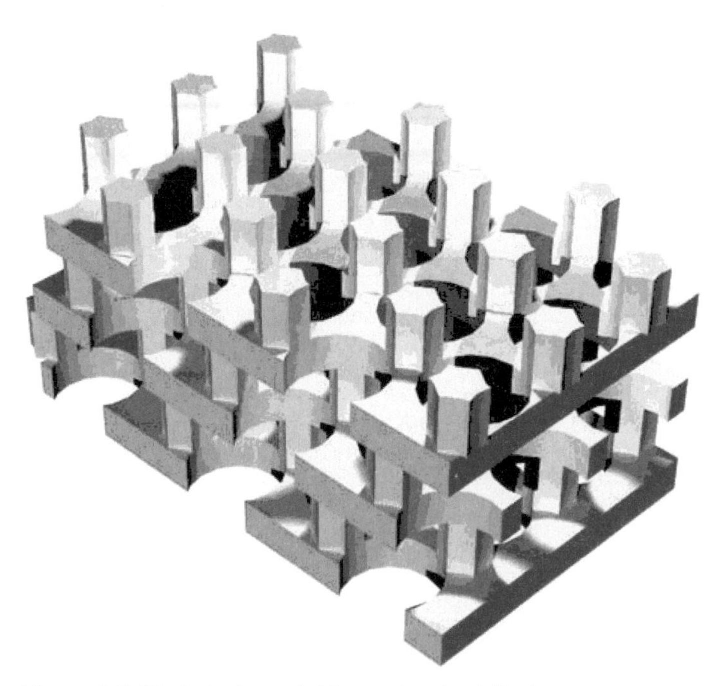

Figure 6.7 3D photonic crystal (computer simulation). The crystal consists of a fcc lattice of air holes in a dielectric. Reprinted with permission from [399]. Copyright 2000 American Institute of Physics.

perfect tunability of the photon properties. For example, with the help of a point defect inside a thin, epitaxially grown film a nanoscaled laser resonator can be fabricated with a typical volume of $(\lambda/2)^3$. This corresponds to just a few tenths of a cubic micrometer.

The realization of photonically useful, periodic structures is in general achieved with the help of micromechanical or lithographical techniques. An elegant alternative is the use of self organization effects. For example, water droplets can serve to generate a three-dimensional hole mask in a polymer film [400]. The method is similar to the mask methods which use ordered arrays of colloidal particles (in general polystyrol- or glass-spheres (see Section 5.2)): in a self organization process an ordered structure of spheres in nanometer size develops; the space between the spheres is filled with a liquid, which subsequently solidifies; following chemical removal of the initial spheres, a skeleton remains which possesses the three-dimensional hole matrix with submicron dimensions. A chemistry-free method has been described in [400]. Here, the ordered hole array is generated with the help of water drops from a stream of humid air colliding with the polymer. Following evaporation of the water

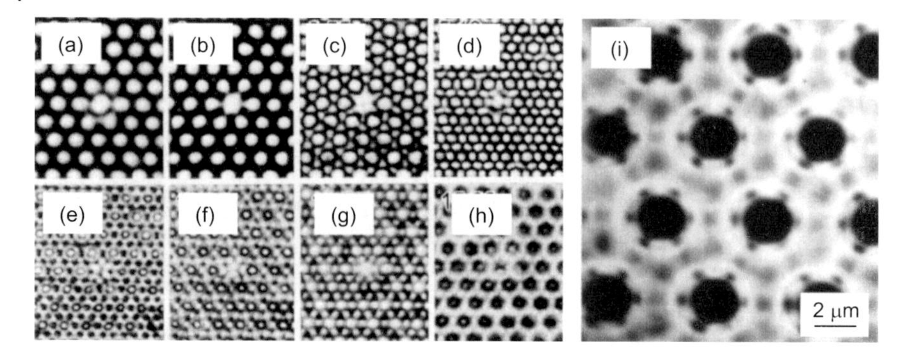

Figure 6.8 Photonic structure generated via self organization of water droplets. Image (i): bright field image of the hexagonally ordered hole distribution in the polymer. Images (a) to (h) are optical cross sections through multilayer hole arrays with (a: −2.55 μm, b: 0 μm, c: 2.55 μm, d: 5.10 μm, e: 7.65 μm, f: 10.20 μm, g: 12.75 μm, h: 15.30 μm). Reprinted with permission from [400]. Copyright 2001 Science.

drops an ordered array remains in the polymer (Figure 6.8). The size of the holes can be varied between 200 nm and 20 μm by variation of the air speed. Figure 6.8 shows hole diameters of about 2 μm.

If one scans the focus plane of a conventional light microscope through such a three-dimensional structure (Figure 6.8a–h), three-dimensional order becomes obvious. For example, the defect in Figure 6.8b in the center of the image vanishes for layers below 10 μm and instead an ordered structure becomes visible.

With diameters of 2 μm the 'artificial atoms' of the photonic crystal from Figure 6.8 are microscopic but not really nanoscopic. In order to obtain nanoscopically ordered arrays, isolated atoms might be periodically captured using light with a predefined lattice distance of the order of the wavelength of the manipulating light. This can be achieved with the help of three-dimensional light lattices, which bind ultracold atoms.

Photonic crystals can be used to fabricate, for example, efficient LEDs, nanoscaled lasers in two-dimensional bandgap materials, efficient crossings in waveguides, selective optical filters, ultrafast optical switches and integrated photonic boards or a new kind of optical fibers ('photonic crystal fiber', PCF). Taken literally, the PCF [401, 402] (Figure 6.9) is not a three-dimensional photonic crystal, but a stretched two-dimensional photonic structure. It consists typically of a hexagonal array of submicron-sized holes, which run parallel over the entire fiber length. These holes concentrate, for example, the guided light onto the central core within the triangular pattern. In other words, the inner total reflection, which determines the guiding of light in a conventional optical

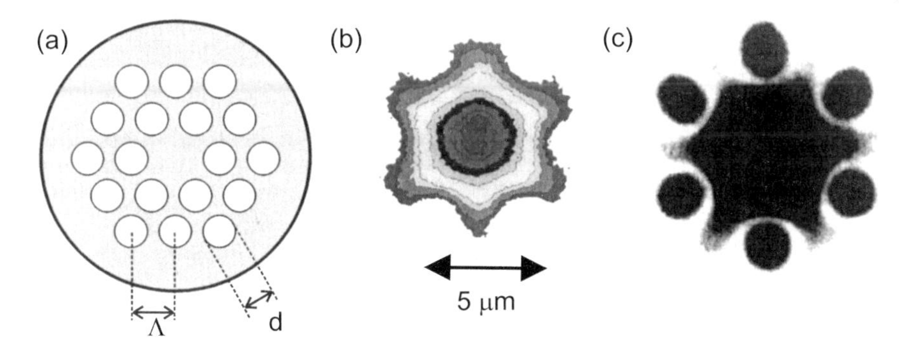

Figure 6.9 (a) Cross section through a photonic fiber. (b) Near field and (c) far field intensity distributions of the transmitted laser light. Reprinted with permission from [402]. Copyright 2001 John Wiley and Sons.

fiber, is modified via an enhanced optical confinement in the special hole array in the fiber cladding. The effective refractive index of the cladding is modified via the insertion of air holes (Figure 6.9). In Figure 6.9b near field and in Figure 6.9c far field intensity distributions of the transmitted light are shown. From the stability and symmetry of the intensity distribution it follows that the photonic fiber acts as a single mode fiber.

Since the light is confined and guided through a central hole within a cladding of photonic bandgap material, an obvious advantage of this kind of light fiber is that they can transport very high light intensities. In conventional fibers nonlinear effects within the inner glass core limit the maximum transportable intensities. However, for many applications within ultraprecise metrology or ultrashort pulse optics, optical fibers with high nonlinearities are also of advantage. In this area PCFs of the kind visualized in Figure 6.9 can be used. The small core diameter results in enhanced nonlinear interaction via self phase modulation while the group velocity dispersion is reduced by an appropriate patterning of the holes. Hence, very short optical pulses can propagate much longer in this kind of fiber compared to conventional optical fibers, resulting in a maximized nonlinear interaction with the glass core [403, 404]. Eventually strongly broadened spectra are obtained which can be used for further temporal compression of the short pulses (cf. the following chapter).

6.1.4
Short- Time Dynamics in Nanostructures

For the last fifteen years or so, ultrashort laser pulses (pico- or femtoseconds, 10^{-12} or 10^{-15} s) have been relatively easily and cheaply available for the generation of structures on surfaces at the micro- or nanometer scale, or for the study of dynamical processes on surfaces with very high temporal resolution [405,406]. One of the main reasons for this opening up of new possibilities is the development of robust, easy to use table top femtosecond laser systems, in many cases on the basis of self-mode coupling, *Kerr-lens mode-locking*, KLM [407].

The absorption of ultrashort laser pulses in liquids or solids results in extremely high electronic excitation rates, which select a few, very fast, relaxation channels from the multitude available. If, for example, the excitation of a nanoparticle in the gas phase proceeds via a multiphoton process, then losses due to stepwise one-photon processes might be avoided. Electronically excited particles fluoresce with lifetimes of the order of nanoseconds. An excitation rate that is four orders of magnitude faster (100 fs) leads, without losses due to fluorescence, to direct ionization. This kind of fast ionization is advantageous, especially if the particle has weak bonds, which would result in strong fragmentation upon conventional ionization via electron impact or nanosecond laser pulses. As a result of the femtosecond excitation the energy is strongly localized, which allows nearly fragmentation-free ionization.

Particles which are adsorbed on surfaces and experience additional loss channels in the form of adsorbate–adsorbate and adsorbate–substrate relaxations, show a similar behavior. Here, femtosecond pulses allow one to obtain fragmentation-free and nonthermal desorption and ionization even in the case of larger molecules up to DNA strands.

As far as structural modifications of metallic or semiconductor surfaces are concerned, purely electronic relaxation processes (for example, electron–electron scattering) proceed on time-scales that are short compared with the pulse length of a typical femtosecond laser. But even in this case at least the coupling to the lattice oscillations via electron–phonon coupling and the subsequent thermal relaxation can be avoided *during* the duration of the laser pulse. More importantly, new (multiphoton) excitation channels exist for all in the high-power regime where the electric field of the laser pulse might exceed the threshold to optical breakdown and ablated material is transformed on an ultrafast time-scale into a plasma. This then shortens the characteristic time-scales for ablation processes, which are otherwise controlled by momentum restrictions due to the relatively large mass of the ablated particles.

The absence of direct coupling to the lattice for laser pulses with durations of less than picoseconds reduces significantly unwanted by-reactions such as vibrational coupling in the case of adsorbates or lattice melting in the case of plain surfaces. However, thermal relaxation processes subsequent to the laser pulse usually make *identification* of the initial laser-induced electronic nonequilibrium distribution at a later temporal stage difficult or impossible. Since, in addition, most optoelectronic measurement techniques are slow (picoseconds) compared with the femtosecond time-scale, the full potential of the temporal resolution becomes available solely in correlation or pump/probe experiments.

In an autocorrelation experiment two light pulses with temporal delay $\Delta\tau$ are frequency doubled in a nonlinear optically active crystal [284]. A Michelson interferometer splits the initial pulse with frequency ω into two pulses, one of them being temporally delayed by increasing the optical path. The crystal is aligned such that phase matching for frequency doubling is only fulfilled if two photons from both beams (separated by a small angle) hit the crystal simultaneously. Behind a color filter light with frequency 2ω is observed as a function of the delay time $\Delta\tau$ between the pulses. Since the intensity of the frequency doubled light is proportional to the square of the incoming light intensity, the observed signal intensity is given by

$$\begin{aligned} I_{2\omega} \quad \propto \quad & \int |[E_0(t)\exp(i(\omega t + \phi)) + \\ & E_0(t-\tau)\exp(i(\omega(t-\tau) + \phi(t-\tau)))]^2|^2 \mathrm{d}t. \end{aligned} \tag{6.1}$$

Besides the frequency doubled signal, which is generated from each of the pulses independently, a time correlated signal is observed, which reaches its maximum if the delay between the pulses approaches zero. A typical measurement for a 51 fs pulse is shown in Figure 6.10.

In the case of an instantaneous nonlinear optical response of the crystal this kind of autocorrelator can be used to determine the temporal pulse shape of the laser. For Fourier limited Gaussian pulses that shape is given by $\tau_{\text{laser}} = \tau_{\text{ac}}/\sqrt{2}$, where τ_{ac} denotes the measured half width of the pulse. If, on the other hand, the pulse width of the laser is well known, one can determine with these kind of measurements characteristic time constants for the generation of a nonlinear optical signal. However, one has to take into account, especially for the measurement of ultrafast processes with time constants below 100 fs, that the light pulse propagates through optical elements, which might disturb it. During such

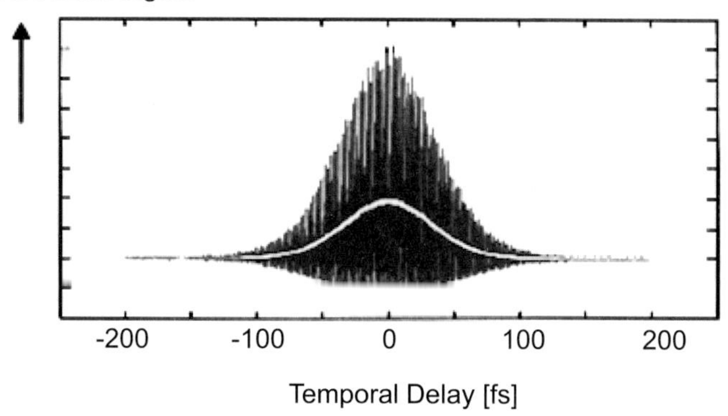

Figure 6.10 Interferometric measured autocorrelation function of second order, measured via frequency doubling of a 51 fs laser pulse. Averaging over the oscillations results in the solid curve, which represents a sech^2 temporal pulse shape [408].

propagation a 'chirp'[4] is generated, which has to be compensated before one draws conclusions about the physical background of observed pulse broadenings or narrowings. It has been shown to be useful for a quantitative judgement concerning the temporal and spectral quality of ultrashort pulses if one measures simultaneously temporal ($\Delta\tau_{\text{laser}}$) and frequency spectra ($\Delta\omega$). In the case of an 'ideal' bandwidth limited pulse with a sech^2 temporal intensity dependence without chirp, the product of the spectral widths $\Delta\omega \times \Delta\tau_{\text{laser}}$=1.978 [409]. Deviations from that value point to a chirp.

In the case of pump/probe experiments one obtains similar results; however, one also has to be equally careful to avoid pulse length changes during propagation through optical elements. Usually the laser pulses have different frequencies, meaning that one pulse can be used for excitation of the system while the second one probes the temporal evolution of the laser-generated non-equilibrium state. In order to maintain coherence, the pulses are generated with the help of additional optical elements (e.g. nonlinear optically active crystals) from the same laser source.

4) With 'chirp' one usually denotes the temporal broadening of the pulse. In the case of normal dispersion ($dn/d\lambda < 0$) the high frequency components are temporally delayed and the low frequency components are accelerated.

Electronic Relaxation at Surfaces and in Ultrathin Films

Besides the fundamental interest on the dynamics of systems with confined dimensions the main goals of time-resolved studies of ultrathin films adsorbed on surfaces are potential applications as ultrafast opto-electronic elements. A large body of literature has concentrated on the dynamic properties of metallic films, especially gold or silver films.

The initial absorption of laser light in thin metallic films results in a collective electronic excitation if the films consist of islands, or appropriate excitation conditions are chosen so as to facilitate surface plasmon excitation . In the latter case the spatial decay of non-localized, that is, propagating, surface plasmons has been determined, resulting in momentum lifetimes of 48 ± 3 fs for a 45 nm thick silver film on a glass prism [410] and of 20 fs for a 40 nm thick Au film [411]. For a 70 nm silver film on a grating structure time-resolved measurements in the ATR geometry [5] have revealed a lifetime of less than 10 fs [412]. The momentum decay times of coherent multiply scattered surface plasmon polaritons in 35 nm thick gold films have also been determined via time-resolved ATR measurements to be about 56 fs with surface roughness leading to a significant increase in damping rate [413]. If the films consist of isolated clusters, then owing to the strong surface scattering the lifetimes of the corresponding localized surface plasmons are expected to be smaller in comparison to those of nonlocalized plasmons.

The decay of surface plasmons means a loss of coherence in the excitation, but there is still a distribution of highly excited, hot electrons. The dynamics of those electrons has been investigated by a variety of transient techniques, taking advantage of ultrashort laser pulses. Besides linear transient reflectivity measurements nonlinear techniques such as transient second harmonic generation (SHG) have proven to be very useful. With laser energies near the interband transition threshold (3.840×10^{-19} J (2.4 eV) in gold) the SH signal is expected to be especially sensitive to transient changes in electron temperatures [414, 415]. In contrast to linear thermoreflectivity measurements, where this change in reflectivity is of the order of 10^{-3} or less [416], changes of the order of a few tens of percent are observed in second-harmonic experiments. Two possible arrangements are shown in Figure 6.11.

In general, the change in reflectivity depends on the transient change in the dielectric function of the investigated metal film, which in turn depends on the density of states, which of course is a function of the local electron temperature. As a consequence, the reflectivity of a metal film that has been excited by a pump photon might be increased or decreased,

5) 'ATR' abbreviates 'attenuated total reflection', see Section 4.3.

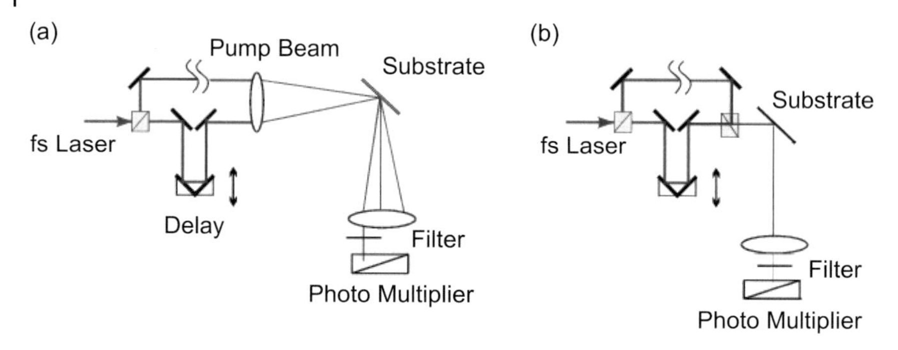

Figure 6.11 Experimental set-up for the time-resolved measurement of transient linear and nonlinear reflectivities. (a) Pump/probe arrangement. The detector monitors the second harmonic signal at 2ω from a probe pulse, while the pump pulse is temporally delayed by changing the optical path in the μm range. (b) Following temporal delay, pump and probe beams are merged again to obtain a collinear autocorrelation function at the sample.

depending on whether the wavelength of the probe photon is above (decreased absorptivity due to an increase in electron occupancy above the Fermi level) or below the interband transition energy [416]. From linear pump/probe experiments on thin noble metal films, delayed thermalization of the electron gas is concluded with typical time constants of 500 fs, but increases into the picosecond regime close to the Fermi energy due to state filling effects that block the relaxation [416]. This electronic thermalization competes with the electron–phonon coupling and makes a quantitative description of the process for metals with strong electron–phonon coupling in terms of the well-known two-temperature model [417] questionable.

The two-temperature model, which is a simple approach to quantitatively describing the thermal response of electrons, to predict reflectivity changes and to obtain electron–phonon coupling constants, assumes a Fermi–Dirac equilibrium distribution for the electrons and a Bose–Einstein distribution for the phonons and thus allows one to decouple the corresponding differential equations. The temporal and spatial evolution of electron temperature T_e and lattice temperature T_l under the influence of a laser source $g(x,t)$ is described one-dimensionally by

$$C_e \frac{\partial}{\partial t} T_e = K_e \nabla^2 T_e - G(T_e - T_l) + g(x,t) \tag{6.2}$$

and

$$C_l \frac{\partial}{\partial t} T_l = G(T_e - T_l), \tag{6.3}$$

which are two coupled nonlinear diffusion equations with electron–phonon coupling constant G, electronic heat capacity $C_e \propto T_e$ and lattice heat capacity C_l, which is constant above the Debye temperature. The electronic thermal conductivity K_e for free electrons is

$$K_e = \frac{2C_e E_F}{3m_e^*(\nu_{ee} + \nu_{el})} \tag{6.4}$$

with Fermi energy E_F and electron–electron, ν_{ee}, and electron–phonon collision frequencies ν_{el} proportional to the lattice temperature.

A model including nonthermal electrons has recently been developed [418], which works especially well under the conditions of low laser power and lattice temperature and yields slower electron–phonon relaxation times compared to Equations (6.2) and (6.3).

An interesting aspect especially for nanoscaled, ultrathin films is the dependence of transient reflectivity on film thickness [419, 420]. With pump pulses at 400 nm the optical penetration depth is 12 nm and is thus significantly smaller than the usual film thickness. However, it has been found that the electron temperature is homogeneous, that is, ballistic electron transport with a velocity of 1000 kms^{-1} dominates for films that are thinner than 100 nm. This results in a significant enhancement of transient reflectivity with decreasing film thickness (Figure 6.12), since the energy that is deposited within the optical penetration depth is ballistically distributed over the film. In addition, the picosecond temporal behavior changes from exponential to linear, in agreement with the prediction of the two-temperature model [420].

It is expected that for ultrathin films (below 16 nm)[6] the exact value of the relaxation time constant depends on the surface morphology, that is, the shapes and sizes of the discontinuous distribution of islands. One then enters the regime of 'supported metal clusters', which is discussed below.

Another spectroscopic possibility to study ultrafast phenomena in thin films or on surfaces that has gained increasing importance within the last few years is time-resolved image potential state spectroscopy, or more general two-photon photoemission (TPPE) [421,422].

The idea is to excite an electron from the conduction band of the solid via a short laser pulse $h\nu_1$, which induces a polarization charge in the metal and thus a Coulomb potential well. This 'image state' has discrete

6) A measure for 'ultrathin' in terms of 'discontinuous' might be the fact that for nominal thicknesses below 16 nm the Au films are charging if irradiated by low-energy electrons, for example, in a LEED apparatus.

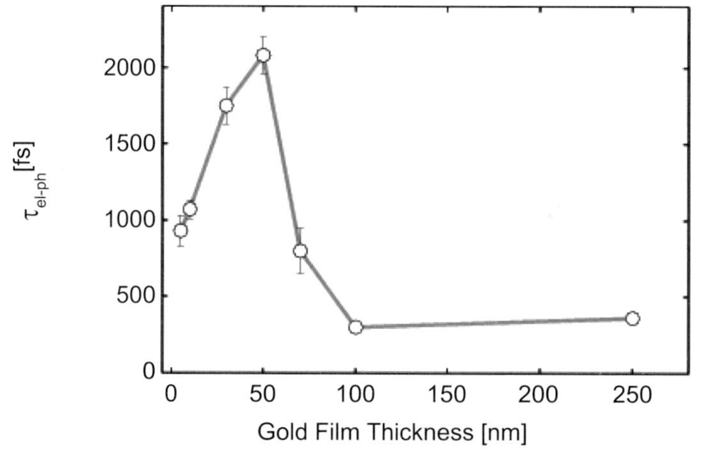

Figure 6.12 Measured electron-phonon decay times τ_{el-ph} for gold films of different thicknesses. Force microscopy images of some films are shown in Figure 6.13.

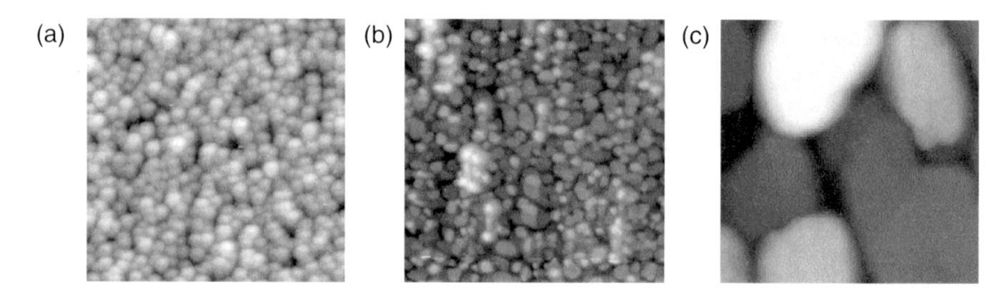

Figure 6.13 AFM images ($1\ \mu m \times 1\ \mu m$) of gold films, deposited onto mica. The statistical roughness of the 5 nm thick film (a) is $\delta = 2.49$ nm, that of the 10 nm thick film (b) $\delta = 3.0$ nm and that of the 50 nm thick film (c) $\delta = 16.03$ nm.

possible binding energies

$$E_B = \frac{-0.85}{n^2} [\text{eV}] \tag{6.5}$$

with respect to the vacuum level V_0, corresponding to a Rydberg series with quantum number n. The accompanying wavefunctions have maxima at 3.17 n^2 above the surface. The image potential state, generated by the first laser, is detected with the second laser pulse $h\nu_2$ via generation of photoelectrons. Since image potential states possess long lifetimes due to their Rydberg character and thus a small line width of a few ten meV, high resolution spectroscopic methods such as two-photon photoemission are necessary in order to detect spectral shifts that are induced by the presence of adsorbates.

The method enables one not only to observe dynamic changes in surface properties such as temperature or adsorbate growth modes, but also electronic changes in buried interfaces or electron-transfer reactions between the surface and adsorbates. The possibility to generate high population densities of image states makes the preparation of a two-dimensional electron gas feasible with nearly ideal, free, mobility, but localization in two dimensions due to the adsorption of adsorbates [423,424].

If one applies lasers with pulse lengths below 100 fs, then a variety of elementary steps in surface dynamics open up for real-time studies [425]. These include radiationless energy relaxation processes of electrons close to metallic surfaces [426, 427], hot electron dynamics [428–432], polarization dynamics on metal surfaces using interferometric TPPE [433] or even electron tunneling processes between STM tips and molecules adsorbed on metal surfaces [434]. The strong coherent coupling between irradiated surface and laser pulse (i.e. the manipulation of the phase) can also be used for coherent control [435] purposes, for example, of photocurrent [436] or photoexcited electron distributions [437].[7] As seen from this, far from complete, list of applications, the main areas of research here are metal surfaces since on those surfaces elementary relaxation processes proceed on a femto- or subfemtosecond time scale. Due to recent developments in laser technology this scale has now become accessible with steadily increasing simplicity.

Relaxation in Nanoparticles

The (nonlinear) optical response of particles with characteristic dimensions of nanometers has been of interest for several decades. Especially in the case of semiconductor nanoparticles this response is due to a size-dependent change in the density of states, which in turn gives rise to large optical nonlinearities. It takes little fantasy to imagine that the size-dependent band gap energies can be used to generate optical (laser) diodes with free-fabricated emission spectra.

On account of the availability of ultrashort laser pulses, the ultrafast dynamic response of such particles has come under renewed interest [49]. Here, as in the case of thin films, one is interested in the exciton dynamics (in semiconductors), the collective and single electron dynamics (in metals) as well as the electron–phonon dynamics [439]. The phonon dynamics of nanoparticles and semiconductor quantum wells

7) Note, however, that coherent control of nuclear motion on surfaces due to strong homogeneous broadening effects (lifetime quenching) [438] is most probably not possible.

has been investigated by Raman scattering. Recent research in the field of cluster physics (i.e. particles with a countable number of atoms) to the phononic properties of large clusters [440] might also find some future applications.

As an example of a nonlinear optical investigation, four-wave mixing studies of cadmium–sulfur–selenide (CdSSe) microcrystallites, embedded in glass matrices, might be mentioned. These studies have shown that the time constant for the generation of a four-wave mixing signal increases with cluster size, whereas the large cross section for this process itself ($\geq 1\,\text{Å}^2$) is nearly independent on size [441]. For small clusters (radius 1 nm) surface-recombination processes are more important than bulk-recombination since the surface-to-bulk ratio is inversely proportional to the radius. Since it is possible to vary the CdSSe cluster size distribution over a wide range, an extremely nonlinear optically active material might be generated from clusters of small radius, the effective time constant of which (given by the charge carrier recombination time) is only about 2 ps.

In the case of metallic nanoparticles, most ultrafast studies have concentrated on the dynamic response of particles in glass matrices (e.g. [442]) or in colloidal suspensions. The case of supported particles is discussed in more detail in the following chapter. In terms of spectroscopy and ultrafast dynamics of colloidal silver and gold nanoparticles, an overview can be found in [443].

Metal Clusters on Surfaces

Discontinuous metal films are of interest for controlling, for example, the polarization in the cover of a waveguide, since their optical properties depend – similar to the case of metal colloids – on the size distribution and average size.

The supported metal islands ('clusters') which form the film, are usually generated usually by thermal deposition of metals on insulator substrates (cf. Section 3.2). As an alternative, one might also generate the aggregates in a cluster aggregation source in the gas phase and deposit them subsequently on the surface (cf. Section 3.2.3). For a given cluster size, a maximum in the absorption probability is observed as a function of excitation energy. In a classic electrodynamic picture, the clusters act as nanoantennas with size-dependent resonance frequencies, combining receiving (absorbing) and transmitting (scattering) properties. Upon irradiating the nanoantenna with an electromagnetic wave, charges are induced at the surface, which result in restoring forces and thus collective oscillations of the conduction electrons.

Within the 'jellium' model the response of the excited electrons is described under the assumption that the ion cores can be represented by a homogeneous, positive background charge. The dielectric response of the solid (the clusters) is, thus, given by the Drude equation

$$\epsilon_2(\omega) = 1 - \frac{\omega_{\mathrm{p}}^2}{\omega(\omega + i\Gamma)} \tag{6.6}$$

with bulk damping constant Γ and bulk plasmon frequency ω_{p},

$$\omega_{\mathrm{p}} = \sqrt{\frac{N_e e^2}{m_e \epsilon_0}}, \tag{6.7}$$

where e denotes the electron charge and m_e the electron mass. The plasmon frequency increases with increasing density N_e of the conduction band electrons since the space charge, induced by the electrons, excites the oscillation. In the case of a cluster, the external field $E(\omega)$ induces a surface charge

$$\sigma(\omega) = \left(\frac{\epsilon(\omega) - 1}{\epsilon(\omega) + 1}\right) \frac{E(\omega)}{2\pi} \tag{6.8}$$

for a metal in vacuum. Obviously, $\sigma(\omega)$ diverges for $\epsilon(\omega) = -1$. Together with Equation 6.6 this means that $\omega_{\mathrm{sp}} = \omega_{\mathrm{p}}/\sqrt{2}$ is a resonance in the surface charge.

This 'surface plasmon resonance' is accompanied by an enhancement of the electromagnetic field strength at the surface of the clusters as illustrated by a Mie calculation in Figure 6.14. Within Mie theory the total light extinction cross section is expanded in a multipole series [444]. The expansion coefficients depend only on the size of the spherical particles and the relative index of refraction of the particle with respect to the embedding medium.

The width and spectral position of this resonance change as a function of cluster size. For clusters with a radius smaller than 1 nm (average number of atoms in the cluster about 100) the resonance shifts with decreasing radius to larger wavelengths [445].[8] The reason for this shift is that with decreasing size the conduction electrons become less strongly bound to the ionic cores. The 'spill out'(Figure 6.15) of the electron density beyond the edge of the cluster increases and the polarizability increases. Hence this effect depends on the cluster material, namely the size-dependent dielectric function. This becomes apparent in Figure 6.15, where electron density distributions are shown for materials

8) This occurs for metals with quasi-free electrons such as sodium and for excitation energies below the interband transition energy.

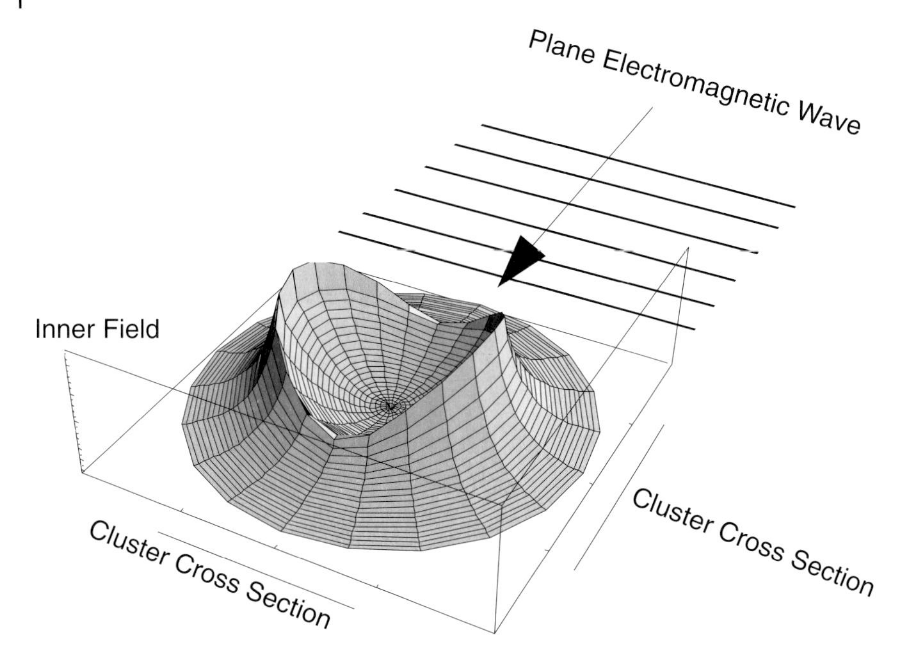

Figure 6.14 Calculated (classic Mie theory [444]) distribution of the modulus of an electric field around a spherical cluster, induced by a plane, linearly polarized electromagnetic wave of wavelength close to a resonance. Note the field enhancement at the surface of the cluster.

with Wigner–Seitz radii of r_S=2 au (corresponding to aluminum) and $r_S = 6$ au (corresponding to an earth alkali metal such as cesium). The electron density is plotted in units of the average density $< n >$, which is related to the Wigner–Seitz radius:

$$r_S = \left(\frac{3}{4\pi < n >}\right)^{1/3} \quad . \tag{6.9}$$

The distance z from the surface is plotted in units of the Fermi wavelength

$$\lambda_F = \frac{2\pi}{k_F} = 2\left(\frac{\pi}{3 < n >}\right)^{1/3} \quad . \tag{6.10}$$

With increasing cluster size the electrons are more strongly bound to the cores and the optical properties can be described, at least along a limited size range (radii of 1 to 10 nm for metallic clusters such as Na_n, cf. Figure 2.3), in dipole approximation using classic electrodynamic Mie theory ([447]). In this regime the spectral position of the dipole resonance is independent of size.

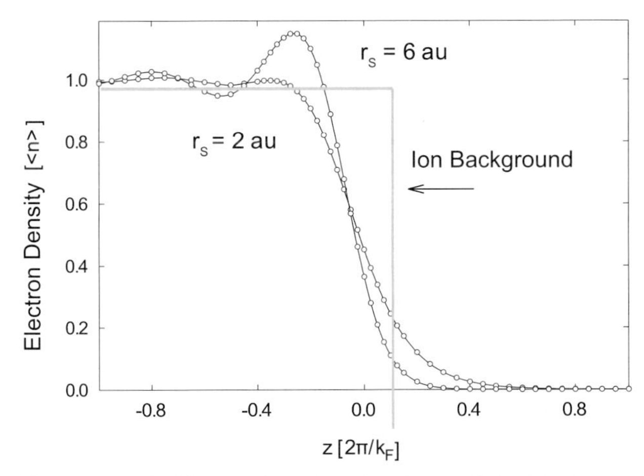

Figure 6.15 Spill-out at the edge of a cluster. Calculated electron density distributions (in units of the average density $< n >$) as a function of the distance z from the surface for $r_S = 2$ au and r_S=6 au [446].

For very large clusters (radii larger than 10 nm)[9] one again observes a red shift of the plasmon resonance. Here, electrodynamic effects such as retardation and excitation of higher order multipole plasmon resonances dominate the spectral position. These latter effects can be satisfactorily reproduced using the (size-independent) optical constants of the bulk material and Mie theory.

Surface adsorbed clusters are, however, not spherical, but ellipsoidal (cf. Figure 6.30). This has a pronounced influence on the position of the dipole resonance as shown in Figure 6.17a. Theoretically the ellipticity of the clusters can be taken into account by expanding Mie theory into 'T matrix' theory [448]. The transfer(T) matrix relates the expansion coefficients of the incoming with those of the scattered electric field and depends (similar to classical Mie theory) alone on the relative index of refraction and the size and shape of the particle. The ellipticity of the particle is taken into account by performing the calculations of the fields within a sphere with radius r_{min} inside and r_{max} outside of the particle (Figure 6.17b). The quality of the method is demonstrated in Figure 6.16, where measured (a) and calculated (b) extinction spectra of sodium clusters on mica surfaces are compared. A clear red shift of the dipole resonance is visible.

9) This 'ionic' radius r corresponds to about $N = 10^5$ atoms, following $N \approx (r/r_{WS})^3$; $r_{WS} = 2.12$ Å is the Wigner–Seitz radius of Na.

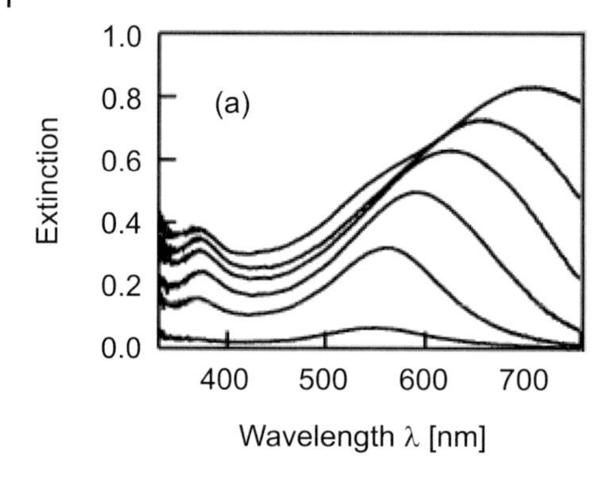

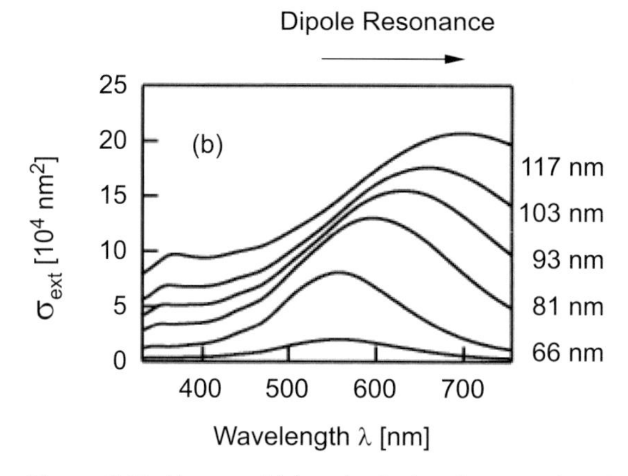

Figure 6.16 Measured (a) and calculated (b) extinction spectra for sodium clusters adsorbed on mica at a surface temperature of 150 K. Between subsequent curves the amount of adsorbed Na increases by a constant value. The calculated curves assume oblate clusters with $R = a/b = 0.5$ with large semiaxes a of the clusters denoted on the right-hand side; b denotes the small semiaxis, and thus R is a measure of the ellipticity of the clusters. For the lowest two theoretical curves the cluster density is $0.8 \times 10^8\,\text{cm}^{-2}$ and $3.2 \times 10^8\,\text{cm}^{-2}$, for the subsequent curves it is $4 \times 10^8\,\text{cm}^{-2}$.

For the calculations one has to assume a size distribution of the clusters, which in analogy to early TEM (transmission electron microscopy) measurements on cold lithium films [449] and gold decoration measurements on insulator surfaces [109,450] is usually written as

$$f_{\pm}(a, a_0) \propto \exp\left[-\frac{(a - a_0)^2}{2\sigma_{\pm}^2}\right], \tag{6.11}$$

with the two widths σ_- and σ_+ related by $\sigma_- = \sqrt{2}\sigma_+$ and the subscripts "+" and "−" respectively denoting cluster semiaxes, a, satisfying the inequalities $a > a_0$ and $a \leq a_0$. As shown in Figure 6.18, this

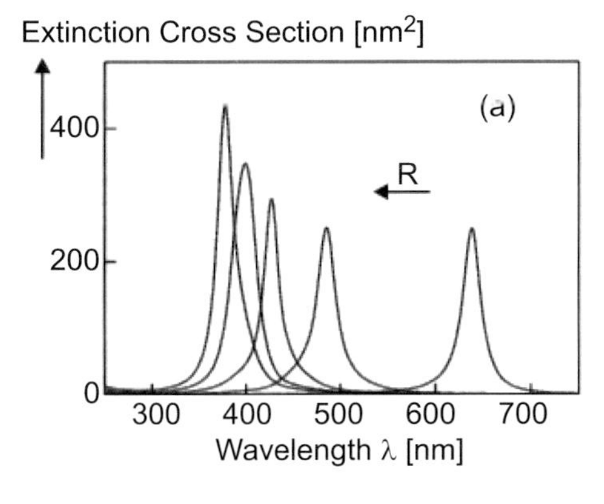

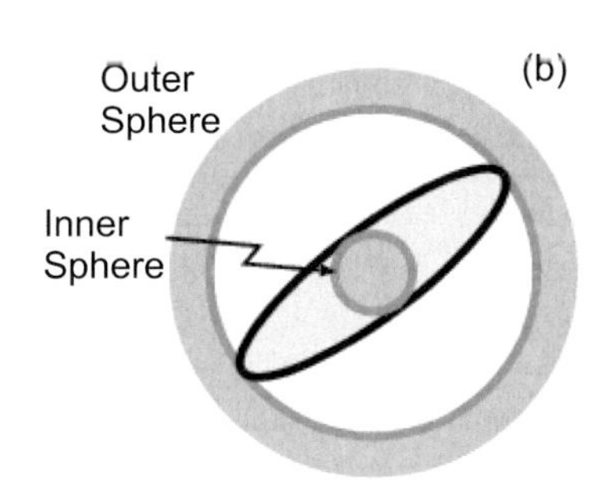

Figure 6.17 (a) Extinction spectra, calculated with the help of T matrix theory for oblate sodium clusters. The large semiaxis has a value of 5 nm, the ellipticity R decreases stepwise in steps of 0.2 from 1 to 0.2 from small to large wavelengths. (b) T matrix theory [287].

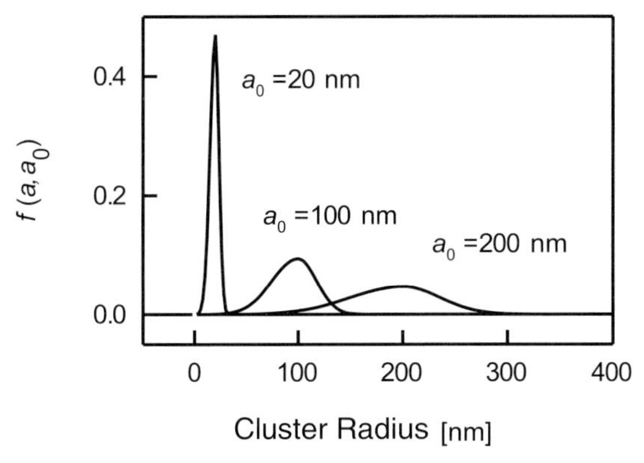

Figure 6.18 Calculated cluster size distributions (FWHM 50% of a_0), using Equation 6.11 and for increasing values of a_0. As can be seen, the distribution outweighs clusters with small radius.

asymmetric distribution is characterized by a FWHM (full width at half maximum) of the order of 50% of the mean cluster radius a_0.

More sophisticated treatments of the optical response of rough cluster films include, the interaction of the clusters with their mirror images in the supporting substrates [451], the mutual cluster–cluster interactions in terms of a quasi-static dipole–dipole approximation [452] and more accurate descriptions of distribution functions of cluster sizes and ellipticities [290]. For example, Figure 4.30a shows the distribution of hydroxylized sodium clusters on a mica substrate. In Figures 6.19 and 6.20 the corresponding distribution functions for small and large semiaxes as

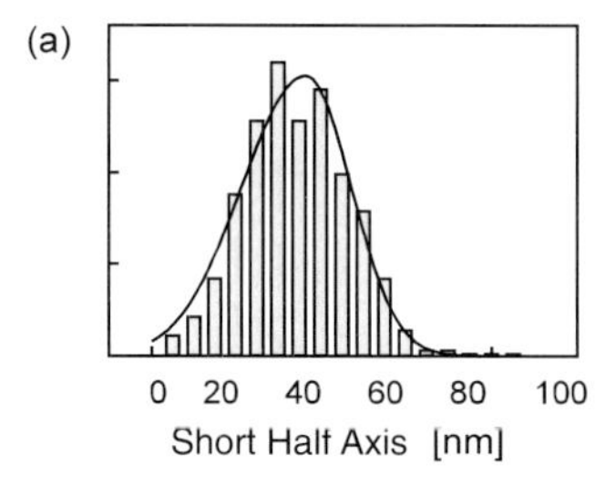

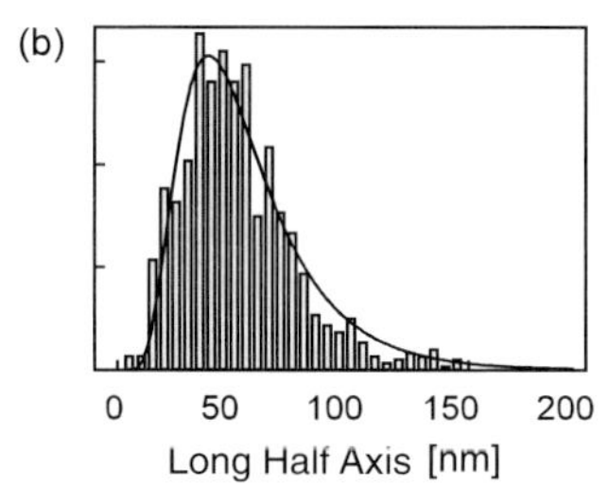

Figure 6.19 Size distributions for the AFM image of NaOH clusters on mica, shown in Figure 4.30a. (a) Small, (b) large semiaxes parallel to the surface. The solid lines are fits assuming a distribution function from Equation 6.11 (a) and a log normal distribution (b).

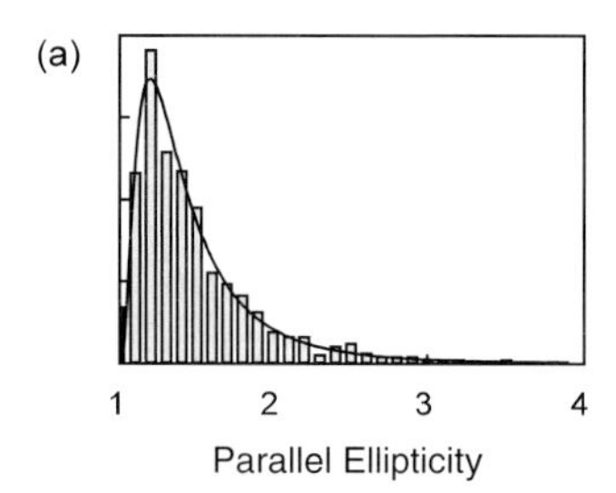

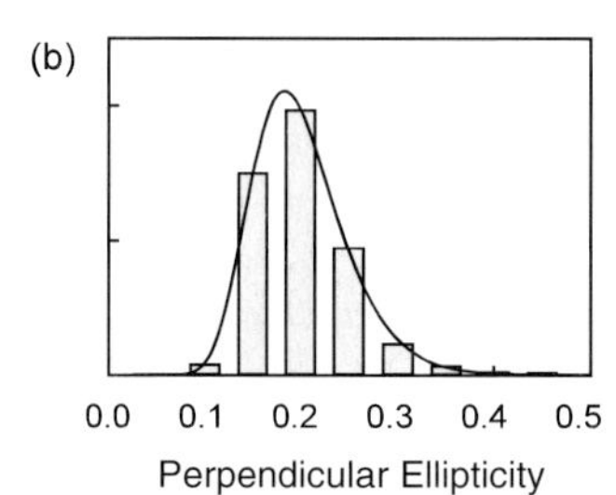

Figure 6.20 Same as Figure 6.19, but distribution of ellipticities parallel (a) and normal (b) to the surface. The solid lines are fits assuming log normal distributions.

well as the ellipticities parallel and normal to the surface plane are plotted.

Obviously, the morphology of thermally grown clusters on a surface is very complex. It has been shown that the observed ellipticities $R = a/b$ can be better described using a log-normal distribution

$$f(R, < R >) \propto \exp\left[-(1/2)\left(\frac{ln(R) - ln(< R >)}{\sigma}\right)^2\right], \tag{6.12}$$

rather than the distribution described in Equation 6.11. The missing agreement between electrodynamic calculated and measured Mie resonance positions might even be used as a tool to deduce specific physical and chemical interface properties [453,454].

Detailed characterization of the morphology of rough cluster films enables one to deduce the electron relaxation dynamics as a function of mean cluster size. As shown schematically in Figure 6.21, upon resonant laser excitation dynamic processes occurring on the femtosecond (initial electronic relaxation), picosecond (coupling to the lattice) and nanosecond (bond breaking processes) time scales are expected.

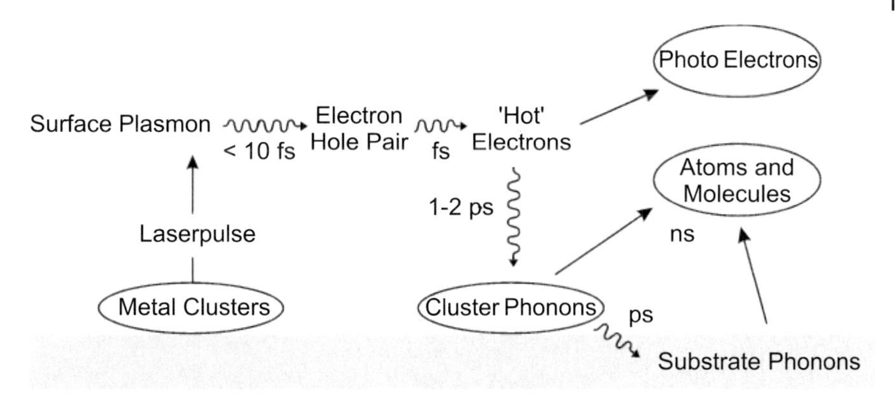

Figure 6.21 Schematic depiction of optical excitation (at $t = 0$) and possible relaxation processes in rough metallic films, consisting of large, surface-bound islands. The ordinate represents the electronic energy E_{electron}, while the abscissa shows typical relaxation times for the decay of collective electronic excitation, single electronic excitation and phononic excitation.

The surface plasmon lifetime is expected to be extremely short (femtoseconds), and it is expected to be size-dependent. For clusters with radii a_0 significantly smaller than the mean free path of the electrons in bulk Na ($\bar{l}$ =34 nm) surface scattering is the dominant damping mechanism in addition to Drude damping. Since the ratio of the surface scattering probability (proportional to the area of the cluster) to the number of scattering electrons (proportional to the volume) scales with $1/a_0$, the plasmon lifetime is expected to be [445]

$$\tau_{\text{sp}} = \left(\frac{v_{\text{F}}}{\bar{l}} + \frac{A v_{\text{F}}}{a_0} \right)^{-1}, \tag{6.13}$$

where the first term refers to Drude damping and the second to surface scattering. Here, v_{F} is the Fermi velocity of the bulk cluster material and A is a size parameter that takes into account electron screening and surface roughness, and varies between 0.38 and $4/\pi$ [445]. For large clusters, retardation effects (radiation damping, excitation of higher multipole plasmons) are expected to broaden the plasmon resonances, that is, shorten the lifetime with increasing a_0 ('extrinsic' or electrodynamic size effects). The total damping rate is then given by

$$\Gamma_{\text{sp}} = \Gamma_{\text{Drude}} + \Gamma_{\text{surface}} + \Gamma_{\text{Mie}}. \tag{6.14}$$

The last term to the size dependence of the lifetime of the dipole resonance can be readily calculated via classic Mie theory. It includes the

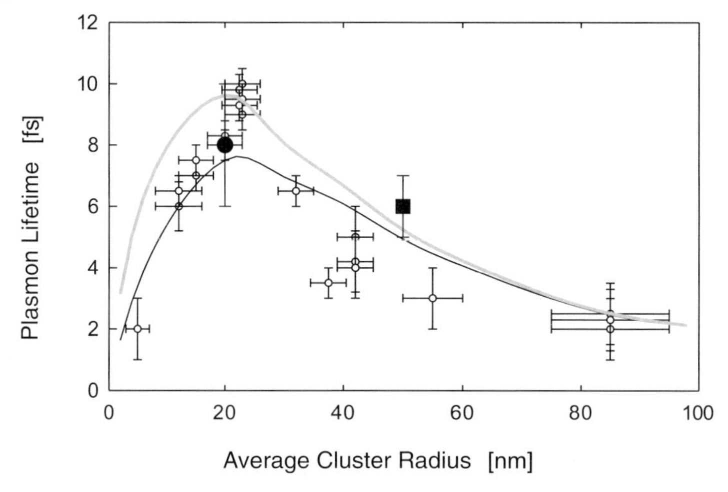

Figure 6.22 Lifetimes of collective electronic excitations in large sodium clusters, measured with the help of ultrashort pulses (open circles [455]). The solid symbols result from measurements on rare metal clusters: silver (circle) [27] and gold (square) [456]. The solid lines represent the classical size effect corresponding to Equation (6.14) for sodium cluster and $A = 1$ (upper curve) as well as $A = 0.45$ (lower curve).

possible effects of interband transitions if the experimentally determined dielectric functions of the bulk are used.

In Figure 6.22 calculated and measured [27,455,456] plasmon lifetimes are shown as a function of cluster size ('classic size effect'), including intrinsic and extrinsic damping mechanisms as well as $A = 0.45$ (upper, gray curve, from density functional calculations for expected values of sodium spheres [28]) and $A = 1$ (lower curve). More recent calculations suggest $A = 0.58$ [31]. For very small clusters this calculation certainly is inadequate [457] and a quantum treatment [458,459] is required. It is to be noted, that this is, even from the optical side, a simplified representation of the interaction mechanism between ultrashort light pulses and nanoscaled particles. Calculations [460,461] and measurements [462,463] show that the initial interaction includes coherent multiplasmon excitation and that the decay can be governed by coupling to multiple single particle excitations. The emission of electrons allows one to use temporally resolved two-photon photoemission for the investigation of the decay dynamics [27,463].

Experimental information concerning the absolute value of the decay time constant for the initial collective excitation of surface bound clusters can be obtained, in principle, both in the frequency domain from the linewidth of the plasmon resonance and directly in the time domain. The spectroscopic method results in a surface plasmon lifetime of, for example, 7 fs for $Na_{n=125}$ clusters adsorbed on BN [464]. However, the width of the surface plasmon resonance is dominated by various homogeneous and inhomogeneous broadening effects such as the cluster size distribution, mutual interactions of the clusters, chemical interface damping [26], and so on, and thus one has to be careful in assigning lifetimes to linewidths. Substantial progress in near-field microscopy has allowed one to measure the homogeneous line shape of *single* gold nanoparticles of about 20 nm radius [465]. From the linewidth a lifetime of about 4 fs has been deduced.

Plasmons in selected metal clusters can also be excited by the tip of scanning tunneling microscope. From the spectral characteristics of the emission of photons following radiative decay a lifetime between 2.2 and 4.7 fs has been deduced for silver clusters between 1 and 6 nm radius deposited onto a thin oxide film [466] and 2.4 fs for gold cluster (3.5 nm radius, height 4 nm) on TiO_2. The latter value corresponds to the value expected for gold clusters in a vacuum [467].

In order to investigate the plasmon lifetimes of larger nano particles, gold particles with 68–260 nm diameter have been lithographically prepared on a transparent tin oxide film such that they have a mutual distance of 10 μm [468]. Plasmons were then excited in the evanescent wave on a prism cathede (cf. Section 4.3). The radiative decay of the plasmons was detected with the help of a microscope objective, focussed on isolated particles. The spectrally resolved light showed clear plasmon resonances. From the linewidth Γ one finds a plasmon lifetime of $\tau_{sp} = T_2/2 = \hbar/\Gamma$ (T_2 is the dephasing time). With increasing particle size the resonance is shifted towards the red spectral range due to increasing radiation damping. In parallel the decay time decreases from 3.6 fs to 2 fs.

The evanescent wave experiment is limited to particles which are larger than 10 nm due to scattered light from surface roughness. Advantages include, the destruction free character, the fact that no laser source is necessary and that the experiment can also be performed in liquids. As usual in evanescent wave or ATR spectroscopy, plasmon excitation is very sensitive to changes in the dielectric environment of the particles – hence the gold particles can be used as nanoscaled sensors of, for example, chemical or biological processes.

For time-resolved measurements of the plasmon lifetime of surface bound clusters, correlation measurements on the basis of nonlinear optical techniques such as optical frequency doubling [455,456,469,470] or two-photon photoemission [27] can be used. Here, too, possible inhomogeneous broadening has to be taken into account [471]. One observes for triangular silver clusters (radius 200 nm) on indium-tin-oxide a lifetime for localized surface plasmons of 10 fs [470]. Strong damping is induced by electron-surface scattering.

Since the second harmonic (SH) signal intensity from the adsorbed clusters is governed by the field enhancement [472] induced by the surface plasmon excitation in the clusters, the duration of this signal provides a direct measure of the plasmon lifetime.[10] This idea, which was introduced by [469], assumes that the laser-induced plasma oscillations can be described by a damped harmonic oscillator which is driven by the femtosecond laser field. By this procedure, a measured collinear autocorrelation function (Figure 6.11) from the adsorbed clusters corresponds to the intensity autocorrelation of the convolution of the temporal shape of the laser pulse (given by a sech^2-function, Figure 6.10) and the exponential decay of the plasmon excitation. In this way, for Na clusters of mean radius 20 nm plasmon lifetimes of the order of 8 fs have been found following excitation near the surface plasmon resonance frequency [455]. A test of the reproducibility of the data was performed by exchanging the Na cluster film for Au thin films, where the SH generation in reflection geometry is instantaneous [473] and where the experimental spectra were found to be unaffected by broadenings.

As a function of mean cluster size, the experiment revealed a pronounced maximum for clusters of average radius 22 nm with lifetimes decreasing for both smaller and larger cluster radii as expected from the classical size effect (cf. Figure 6.22). For large cluster radii one expects that the simple theory is not able to reproduce the experimental results since it does not take into account cluster–cluster interactions. These interactions influence the local fields for distances between neighboring clusters smaller than four times their radius [452,474]. This would be the case for clusters with $a_0 \geq 34\,\text{nm}$.

The decay of the excited surface plasmons results in the distribution of hot electrons. The thermalization time constant of this distribution can be estimated as the electron–electron collision time constant, which can be calculated from Fermi liquid theory to be [475]

$$\tau_{\text{ee}} = \frac{128}{\pi^2\sqrt{3}} \left(\frac{E_{\text{F}}}{E_{\text{laser}} - E_{\text{F}}} \right)^2 \frac{1}{\omega_{\text{p}}}. \tag{6.15}$$

10) There is always a certain probability of SHG due to the nonzero value of $\chi^{(2)}$. However, this signal would be too small to be measured without field enhancement.

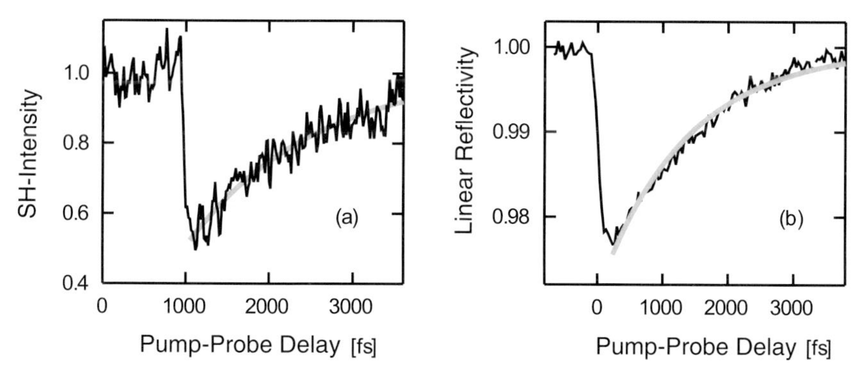

Figure 6.23 Nonlinear (a) and linear (b) reflectivity changes as a function of pump/probe time delay. (a) Na/LiF, $\tau = 1.1(1)$ ps. (b) 10 nm Au/mica, $\tau = 1.07(6)$ ps (J.-H. Klein-Wiele et al., private communication, 1998).

With $\hbar\omega_p = 8.960 \times 10^{-19}$ J (5.6 eV) and $E_F = 4.992 \times 10^{-19}$ J (3.12 eV) for bulk Na, one obtains $\tau_{ee} \approx 10$ fs.[11] Hence, it is likely that during a laser pulse of several tens of femtoseconds length local electronic equilibrium is obtained. The subsequent thermalization time of the phonon subsystem, τ_{ph}, is given by the ratio between the phonon mean free path and the speed of sound in bulk sodium. Again, for Na one finds $\lambda_{ph} \approx$ 1500 Å at 150 K as deduced from the lattice constant of 4.28 Å and $c_s =$ 4340 ms^{-1}. Thus $\tau_{ph} \approx 35$ ps which is three orders of magnitude larger than the thermalization time of the electrons and a factor of 20 larger than the electron–phonon loss time, a measurement of which is described below. Hence, the lattice is not expected to be in equilibrium during the whole period of measurement.

The subsequent picosecond dynamics of the highly excited clusters can be investigated by time-resolved pump/probe measurements, just as in the case of ultrathin metal films (see Section 6.1.4). The electronic excitation by a first (pump) laser pulse and the corresponding increase in electron temperature results in unoccupied states below the Fermi edge and enhances the absorbance of the second (probe) pulse. It thus leads to a decrease in SHG from the probe pulse (Figure 6.23).

The change in electron density due to the femtosecond pulse, and the Fermi edge-smearing associated with the heating of the conduction band electrons affects both the second-order nonlinear susceptibility and the linear dielectric function via the Fresnel factors for ω and 2ω. Both from

11) This extremely short time constant for interelectronic collisions is due to the high frequencies accompanied by the strong laser excitation. It has been noted, however, that in the case of strong laser excitation τ_{ee} might increase due to a cascading of the hot electrons [429].

experiments on polycrystalline Ag and Au surfaces [414] as well as from phenomenological theory [415] it is concluded that the electron temperature dependence of $\chi^{(2)}_{zzz}$ determines the picosecond time response. A similar behavior is expected for alkali cluster films.

As the hot electron distribution cools down owing to collisions with the lattice of the clusters, the SHG from the probe pulse recovers. Thus, by detecting the SH signal from the probe beam as a function of pump/probe delay the electron–phonon time constants of the cluster films are directly measured. For Na clusters of radii between 30 nm and 50 nm, adsorbed on lithium fluoride, the values of the temporal half width are $\tau_{ep} \approx 1$ ps [476,477] (Figure 6.23), which is similar to the values found on thin films of noble metals (Figure 6.12)[478].

Finally, a closer inspection of Figure 6.23 shows that for long delay times the second harmonic signal (as well as the linear reflection signal) does not completely reach the initial intensity level. This is attributed to phonons returning energy back to the electron gas and thus keeping the electron temperature high over a longer temporal period. The fraction of phonons participating in this effect has been estimated to be of the order of 10^{-3} [479].

Picosecond Electron Diffraction

Up to now we have discussed methods to study – in real time – vibrational and electronic energy transfer on surfaces by the use of ultrafast laser spectroscopic techniques which might be called 'conventional' in the sense that they could also be (and have been [480]) applied to problems appearing in the gas or liquid phase. Some indirect information on structural changes of surfaces could be deduced from the sensitivity of second harmonic generation on surface symmetry. However, it would be rewarding to improve the temporal resolution of conventional surface structural probes such as electron diffraction and in that way monitor microscopic changes of surface morphology in real time. This would ultimately include watching the vibrational motion of surface atoms, the hindered rotational motion of adsorbed molecules [481] or the evolution of elementary steps of surface reactions.

Real space methods involving ultrashort laser pulses and scanning tunneling microscopy such as 'correlated optical reactivity and STM, CORSTM' [482,483] are just evolving,[12] while scattering methods have a long history.

12) If one is interested in macroscopic structural information on laser-induced surface morphology changes, then conventional light microscopy might be combined with ultrashort pump/probe techniques. See [484].

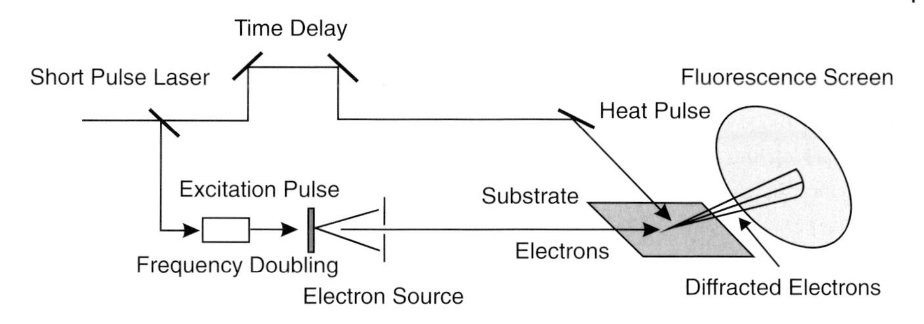

Figure 6.24 Experimental set-up for the time-resolved measurement of surface structural changes via reflection electron diffraction.

An intriguing idea for picosecond electron diffraction is to employ a streak camera [485], which upon irradiating the photocathode with an ultrashort laser pulse generates a photoelectron replica with a temporal width of picoseconds and spatial resolution of several tens of micrometers [486]. By synchronizing the electron pulse with a heating laser pulse the temporal evolution of heat on a surface can be monitored by means of the transient surface Debye–Waller effect (i.e. intensity changes of reflection high-energy electron diffraction (RHEED) peaks) [486, 487]. A typical set-up is sketched in Figure 6.24.

The first time-resolved structural studies have been performed on model surfaces (Pb(110), Pb(111) and Pb(100)), on which the phenomenon of surface melting has been investigated extensively in the past. In contrast to bulk melting the surface melts by forming a disordered layer, several atom layers thick, at temperatures significantly below the bulk melting point [488]. Heating the surface by a short-pulse laser and watching the transient structural changes by RHEED has revealed that the surface disordering occurs on a time-scale below 180 ps, that the surface disorder temperature in the case of Pb(110) is the same as in the static case even if the heating rate is of the order of 10^{11} Ks^{-1} and that the process is reversible, that is, the crystalline order can reform [489]. In contrast, on the more close-packed Pb(111) surface the high heating and cooling rates allowed superheating up to 120 K before melting started [490]; that is, the disorder temperature was seen to be significantly above the bulk melting point. Finally, in the case of Pb(100) one observes incomplete surface melting with finite disordered layer thickness and residual order up to the bulk melting temperature and above [491]. These studies are especially important since there is a strong interplay between surface melting and surface roughening. Reversible or irreversible roughening is of cardinal importance for such diverse phenomena as growth of thin films or annealing of sputtered surfaces.

Even more detailed information can be gleaned if one applies picosecond RHEED and laser heating to different growth phases of adsorbates on surfaces, where, for example, the typical RHEED growth oscillations deliver additional structural information on the ultrathin films. Also, very recently cross-correlation measurements combining femtosecond laser pulses (λ =800 nm) with picosecond X-ray pulses have provided information about ultrafast structural disordering processes, at least in the bulk of an InSb crystal [492]. Extension of these measurements to the study of surface processes might become feasible in the near future.

Experiments using 100 ps temporal resolution suggest that for the Ge(111)-c(2x8) - (1x1) reconstruction in the case of laser heating disorder (i.e. a phase transition) only begins at 584 K, while for thermal heating a phase transition occurs at 510 K [493]. In the case of the high temperature phase transition above 1000 K and laser heating, a super heating of the upper double layer occurs and the phase transition runs at about 60 K higher temperatures than compared to thermal heating [494]. The uppermost double layer expands perpendicular to the Ge(111) surface a factor of 11 more than the bulk in the temperature range between 300 K and 600 K [495].

Finally, the picosecond RHEED method is also used to obtain information about ultrafast changes of structural properties of gas phase particles [496]. In addition, structure and dynamics of molecules in the course of a reaction can be observed [497].

6.2 Electronics

6.2.1 Optoelectronics

The attachment of functional groups on self organized or via the Langmuir–Blodgett technique generated ultrathin organic films (see Section 3.2.2) allows one in principle to fabricate a multitude of nanoscaled optoelectronic elements. For example, a variable degree of optical anisotropy can be achieved via attaching functional side groups, resulting in birefringent ultrathin films. As a result, a linear optical filter is obtained the properties of which can be changed by applying an external electrical field that orients the side groups.

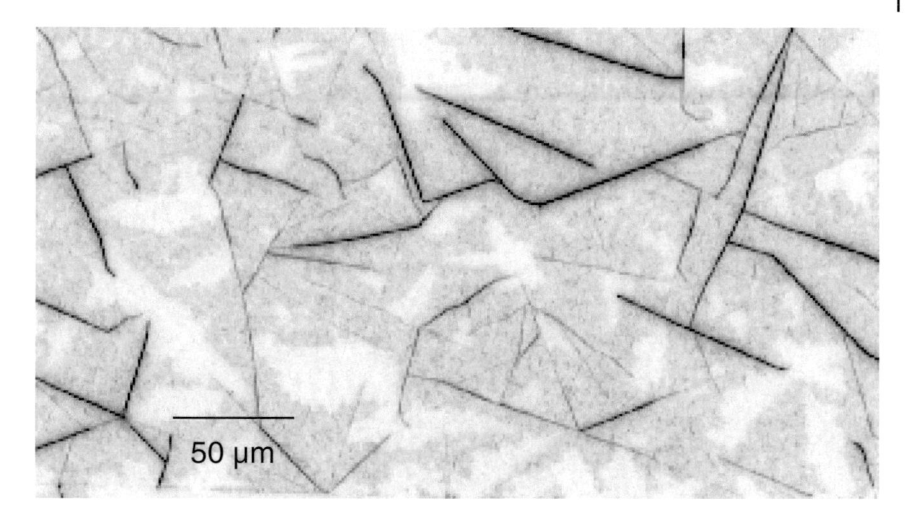

Figure 6.25 Fluorescence microscopy image of light emitting, needle-like aggregates from organic molecules (p-5P), which decorate defects on a sodium chloride surface.

If one uses organic molecules with delocalized π-electrons which are optically active even without chemical modification (e.g. phenylene oligomeres as discussed in Section 4.2.2), nanoscaled connections between elements on surfaces can be generated. Here it is possible to take advantage of the fact that growth of organic aggregates on dielectric surfaces occurs preferentially at sites with increased binding energy, that is, defects, which even steer the direction of anisotropic aggregates such as needles. Phenylene-oligomer needles,for example, decorate the edges and corners of alkalihalide crystals (Figure 6.25).

On smooth dielectric surfaces without specific defects the growth of the aggregates is dominated by the orientation of microscopic surface dipoles (Figure 3.15). Morphological and optical properties strongly depend on the microscopic structure and the growth conditions. If those are known and controlled a macroscopic amount of nanoscopic, well defined aggregates with optoelectronic activity can be fabricated. A connection between the components can be achieved,for example, via submicron-sized optical waveguides, demonstrated in Figure 3.17 [13].

Optical data transport via waveguides (at present of course mainly on a macroscopic scale) is one of the most important growth areas of information technology. The bandwidth of information that is transferred

13) The propagation losses in these nano-waveguides for visible light is, however, rather high because of the low confinement and absorption in the materials used.

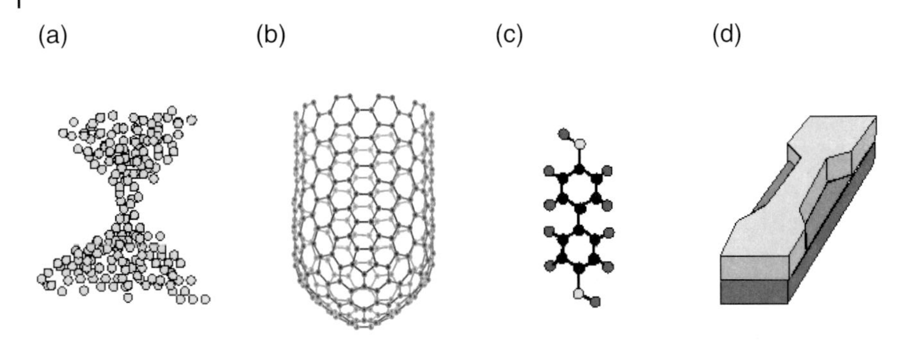

Figure 6.26 Nanoscaled electric conductors: (a) Chain from individual metal atoms; (b) carbon nanotube; (c) molecular wire; (d) semiconductor quantum wire.

currently doubles more than twice as fast as the performance power of electronic components. Hence there is a huge need to develop new light sources (multi wavelength lasers), new optoelectronic elements (improved multiplexer, WDM, FWDM, splitter etc.) and lossless waveguides.

6.2.2
Molecular and Nanoelectronics

Wires

Molecular nanoelectronics dates back to predictions from 1974, where the advantages of using molecules as diodes between two electrodes were discussed [498].

An important prerequisite for useful electronics on the nanometer scale is electric conduction connections in the submicron range. Quantum wires made from semiconducting materials such as GaAs/AlGaAs, have been thoroughly investigated in this context. In many cases these have been produced with the help of high resolution lithography (cf. Section 2.3.1).

Carbon nanotubes can also serve as efficient conductors. The advantage of these conductors are high mechanical stability and a well defined diameter,which ranges from a nanometer (for a tube with a single wall) to a few ten nanometers (for tubes with multiwalls). These nanotubes are most easily produced via carbon plasmas.

Electric wires with dimensions of a nanometer or below can be, in principle, molecular wires, made from carbon, nitrogen, hydrogen and sulfur atoms or wires from individual gold or silver atoms. In the latter case we talk about one-dimensional conductors, the conductivity of which is close to the quantum value of conductivity [499]

$$G_0 = \frac{2e^2}{h} \approx 8 \cdot 10^{-5} S\,. \tag{6.16}$$

The most intensively investigated molecular wires made from organic molecules are based either on n-alkane chains ($CH_3-(CH_2)n-1$) or on conjugated polymers such as poly-paraphenylenes. N-alkanes have a large electronic gap between the HOMO (highest occupied molecular orbital) and the LUMO (lowest unoccupied molecular orbital) of about 9.600×10^{-19} J (6 eV). Thus they are nearly insulating. Phenylenes behave with band gaps of around 3.200×10^{-19} J (2 eV) to 6.400×10^{-19} J (4 eV) and delocalized π-electrons more like semiconductors. Both classes of molecules can form well ordered self-organized films (SAMs) on the surface of gold, if they are functionalized with thiole (SH) or isocyanide (CN) end groups.

The current–voltage curve for a molecular wire can be described approximately via

$$I \approx \frac{2e}{h} \int_{\mu_1}^{\mu_2} T(E,V)dE \quad , \tag{6.17}$$

where μ_i are the electrochemical potentials of the contacts (e.g. surface and STM tip) and $T(E,V)$ is the transmission function of the molecule, which follows from the molecular energy levels and their coupling to the substrate. Via differentiation of Equation 6.17 the conductivity of the molecular wire follows:

$$G(V) \approx \frac{2e^2}{h} 0.5 \cdot [T(\mu_1) + T(\mu_2)] \quad . \tag{6.18}$$

The quantum unit for the conductivity (Equation 6.16) is thus modified via the symmetric averaging over transmission functions with respect to the electrochemical potentials of the contra points.

Figure 6.27b shows the conductivity for a molecular phenlyene thiole wire on a gold surface, measured with the help of a scanning tunneling microscope [500]. The band gap of about 6.400×10^{-19} J (4 eV) as well as the good agreement between measurements (points) and calculation (solid line) are visible. Negative voltage means that electrons are moving from the gold surface into the tunnel tip. For positive voltage the electrons float from the tip into the surface. This means for the molecular wire that electron conduction occurs via the highest occupied molecular orbital (HOMO) if the Fermi level is closer to the HOMO as compared to the lowest unoccupied molecular level (LUMO) or vice versa.

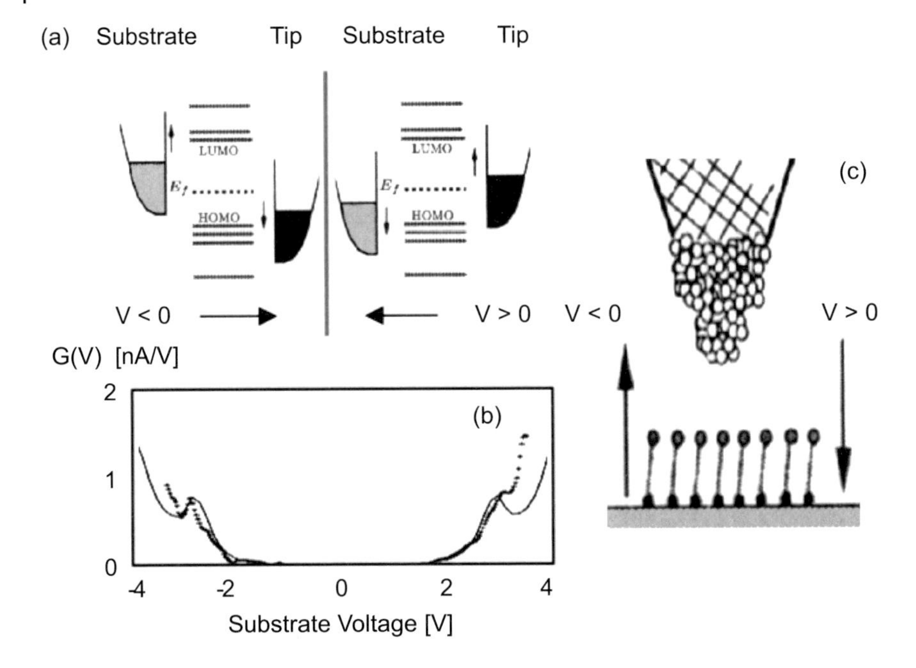

Figure 6.27 Conductivity measurement on molecular wires. (a) is the term scheme for negative and positive bias of the STM and (b) is a typical conductivity measurement for phenylene thioles. (c) a tunnel microscope tip is shown schematically above a self organized thiole film on a gold surface. Reprinted with permission from [500]. Copyright 2000 Academic Press.

Systematic measurements as a function of wire length show that for molecules with thiol end groups the band gap decreases with an increasing number of phenylene rings – as one would expect from the increasing number of delocalized electrons. Interestingly, phenylene wires, functionalized with isocyanide end groups show the opposite effect. Apparently the current through a molecular wire does not only depend on the wire itself, but also to a large extent on the end group binding to the surface[14]. This is true to a much lesser extent for isolated, conducting polymer wires from, for example, polyaniline [501].

Instead of taking advantage of the electric conduction directly via the organic wire, the wire can also be used as a template for the generation of conducting wires with dimensions in the nanometer range [502]. For

14) Similar observations have been made regarding the mechanical stability of nanowires[13].

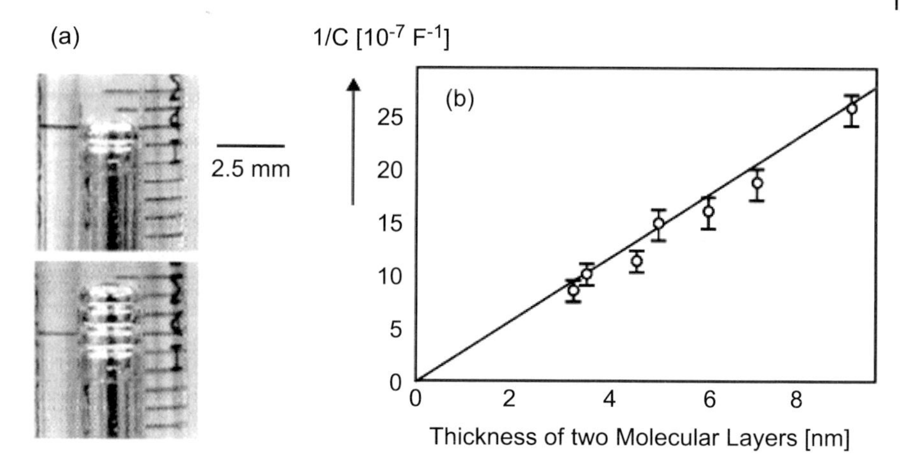

Figure 6.28 Chain-length dependence of the inverse capacity of a Hg-SAM/SAM-Hg capacitor. (a) Photographs of the Hg-SAM droplet system within the syringe are shown. Reprinted with permission from [503]. Copyright 1998 American Institute of Physics.

this purpose DNA has been proven to be an especially appropriate bio molecule; see also Figures 6.38 and 6.39.

Nanocapacitors

In order to attach a metallic film on top of the organic film as an element of a well defined nanoscaled structure, diffusion of the metal into the organic film has to be avoided (Figure 6.29). For the nanocapacitor element shown in Figure 6.28 this problem has been solved by mutually contacting two liquid mercury droplets in an ethanolic alkane thiole solution inside a syringe. Because the SAMs are adsorbed with perpendicularly oriented molecules on the liquid mercury droplets, no domains are generated. Thus, there are also no domain borders at which diffusion through the organic film can occur. The thickness d of the organic spacer layer increases proportional to the chain length. The SAMs adsorb perpendicular to the surface since the tilt angle of the molecular axes with respect to the surface normal (cf. Figure 3.14) is determined by the interaction between possible adsorption sites and spatial requirements as given by the van der Waals radii of the organic molecular chains. For a liquid surface the boundary condition of possible adsorption sites is no longer applicable.

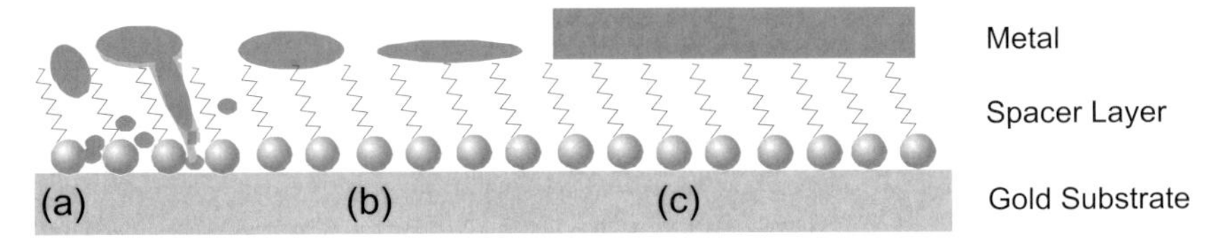

Figure 6.29 Growth modes of metals on ultrathin organic films: diffusion (a), cluster formation (b) and film formation (c).

The capacity of the nanocapacitor is given similar to a macroscopic capacitor as

$$C = \epsilon_0 \epsilon \frac{A}{d}, \tag{6.19}$$

where A denotes the area of the capacitor. Hence one expects with increasing chain length a linear increase in $1/C$ if the dielectric constant ϵ of the SAMs is independent of chain length. As seen from Figure 6.28, this is obviously the case. One obtains $\epsilon = 2.7 \pm 0.3$, a value that agrees with the one found for alkane thiols adsorbed on gold (ϵ=2.6). The measured value of capacity is, for example, for CH_3—$(CH_2)_{17}$ – SH, $C = 8.6\,\text{nF}$.

Of course the range of possible applications would be much larger if one could fabricate such nanostructured elements on solid surfaces. However, i) defects within the organic films, at which current leakage might occur, and ii) diffusion of the metal atoms from the electrons into the organic film, mean that such is problematic.

Figure 6.29 shows schematically possible growth processes of metallic structures on ultrathin organic films. If one evaporates the metals at room temperature directly onto the organic films, there is a high probability that they will diffuse in between the molecular chains (Figure 6.29a). Systematic investigations have shown that the diffusion probability is inversely proportional to the reactivity of the metals on the surface of the organic film [152]. This means that silver atoms have the smallest possibility of forming a well organized metallic film on SAMs, while this probability increases in the order copper, nickel, potassium, sodium, aluminum, chromium and titanium. In an experiment on the electric rectification on a nanoscaled system similar to the one shown in Figure 3.14, a titanium film has been successfully used as a defined upper metallic electrode [504].

The diffusional motion inside the organic films can be significantly suppressed by cooling of the sample. Disadvantageous is that on the cold organic film surface the possibility to form a homogeneous metallic

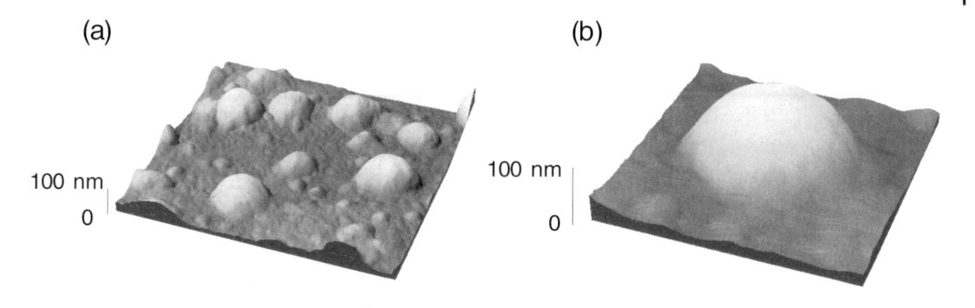

Figure 6.30 Atomic force microscopy images of alkali clusters on top of a dodecane thiole coated gold film. Image sizes are $(950 \times 950)\,\mathrm{nm}^2$ (a) and $(300 \times 300)\,\mathrm{nm}^2$ (b).

film is very low. Instead, the preferred growth mode will be Volmer–Weber growth (Section 3.2.1), that is, the generation of very rough films with large clusters. More recent approaches use electrochemical methods or post-polymerization of SAMs.

Quantum Dots in Layered Systems

Figure 6.30 is an atomic force microscopy image of an ultrathin organic film (dodecane thiole) adsorbed on a gold sputtered mica crystal. At low temperatures (150 K) alkali atoms have been evaporated onto the organic film. The force microscope is rather insensitive to the organic film and mainly images the alkali aggregates. As seen, these consist of large, separated clusters, which have the shape of flat ellipsoids. Such discontinuous alkali films are of general interest for nanooptical and nanoelectronical applications since they show wavelength selective field enhancement effects (surface plasmon excitation, Section 6.1.4), have ultrashort response time and represent isolated conducting islands on an insulating barrier layer.

Important for possible applications is that the metal clusters are electronically separated from the metallic support (the gold surface) via an organic film, that is, they have no direct contact. Whether this condition is fulfilled or not cannot be determined via atomic force microscopy measurements. As an alternative, it is possible to check whether the linear and nonlinear spectra of the clusters are influenced by the presence of the gold film. It is expected that the dipole resonance which dominates the optical spectrum of the clusters shifts as a function of distance between cluster and gold film as induced dipoles, and thus image charges, are induced in the gold film. This has been experimentally proven [505]. Is the organic film sufficiently thick (a few nanometers), resonance shifts and

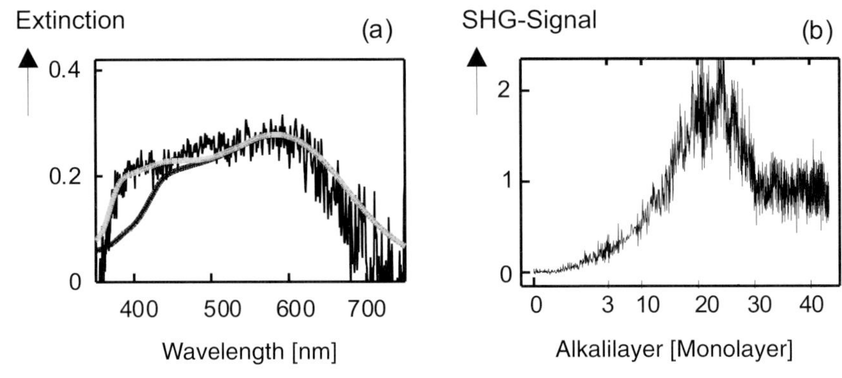

Figure 6.31 Linear (a) and nonlinear optics (b) of the alkali clusters shown in Figure 6.30 [507]. (a) Measured and theoretically fitted extinction spectrum. (b) Optical frequency doubling at alkali clusters.

dampings vanish and one obtains the linear [506] and nonlinear spectra [507] of the free clusters.

Figure 6.31a shows an extinction spectrum of the sample from Figure 6.30, together with two calculated curves. The grey curve results from T-matrix calculations, for which the size distribution of the clusters has been fitted to the measured values. The black curve results from a direct topographical analysis of the cluster distribution as determined via atomic force microscopy. For large wavelengths the curves agree very well. The discrepancy at short wavelengths results from the fact that due to the finite size of the tip, atomic force microscopy does not correctly image the small clusters with diameters smaller than 10 nm.

In Figure 6.31b the second harmonic (SH) activity of $Na/C_{12}/Au/mica$ in reflection is shown as a function of increasing coverage after excitation with 1064 nm light. Without alkali metal no SH activity is visible due to the small electromagnetic field strengths. With increasing alkali metal size the SH signal increases strongly due to surface plasmon enhancement in the clusters [508]. Behind the plasmon resonance it decreases to a constant value. The presence of the organic film apparently does not disturb the frequency doubling efficiency.

While alkali metals have large advantages in terms of optical properties due to their nearly free outer electrons and their high oscillator strengths, their disadvantage in terms of nanooptics is their reactivity. Hence, one has to protect the surface against reactive gases. Films from self organized organic molecules (e.g. alkane thiols) are very useful for that purpose.

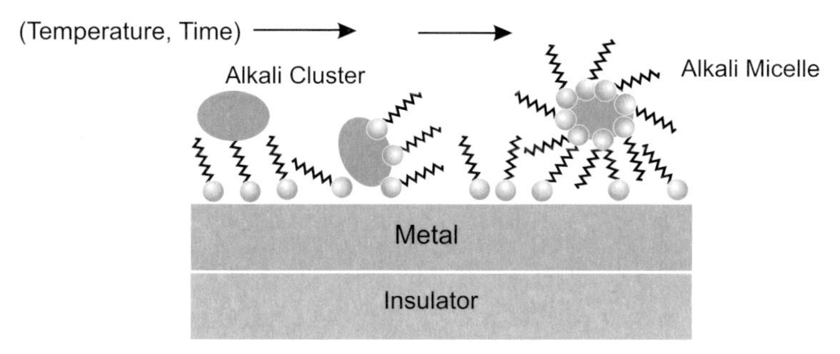

Figure 6.32 Model for micelle formation on alkane thiol monolayers following heating from 150 K to 300 K.

A possible approach is shown schematically in Figure 6.32. If one prepares the metallic clusters at low temperatures on a monomolecular alkane thiole film and subsequently heat the substrate to room temperature, the organic molecules will be desorbed from the gold surface and they will form a protecting layer around the alkali clusters. It has been shown that the optical properties of the metal clusters are conserved below this protecting coating even if one applies ambient air to them [507].

Instead of growing the metal nanoparticles directly on the organic spacer layer, the metal clusters can also be prepared in colloidal solution. The organic film coated gold crystal is dipped into the cluster solution, and the clusters are deposited onto the organic film. On top of the metal [333] or semiconductor cluster layer [509] further layers of organic film and metal clusters from solution can be deposited. Eventually a three-dimensional super lattice of metal or semiconductor quantum dots is generated.

Single Electron Transistors

Owing to the possibilities to structure them, for example, with a scanning tunneling microscope [510], ultrathin films from organic molecules should be useful for fabricating single electron transistors (SET) similar to semiconductor quantum dots.

A SET circuit consists of a capacitor for charge storage and a very thin barrier layer, which allows single electrons to tunnel as discrete charge packets between two electrodes. If a metallic island is situated between two tunnel connections, the electric potential of that island depends strongly on the number of additional electrons on the island. As a result the transition of additional electrons is blocked until a threshold of the applied voltage is overcome ('Coulomb blockade', Figure 6.39). The potential of the charged island and thus the blockade range can be varied in a quantized fashion by applying a voltage on an additional gate electrode.

The single electron effect becomes visible once the charge energy $e^2/2C$ is much larger than the thermal energy k_BT and the energy of quantum fluctuations. Hence the capacity of the island C has to be small enough and the resistance of the tunnel connections large compared to $h/2e^2$ (the quantum unit of resistance). Is the resistance too small, the quantum mechanical uncertainty of the position of the charge quanta on the electrodes results in statistical tunneling effects, which destroy the expected transistor property.

Typical minimized tunnel contacts have areas of $60 \times 60\,\text{nm}^2$ and total capacities C of a few 10^{-16} F at a resistance of $100\,\text{k}\Omega$ per element. In order to suppress thermal effects usually temperatures below 0.1 K are necessary.

In the organic SET, which could work at much higher temperatures, a gold film would be used as ground electrode on top of which an organic film is grown as barrier layer, followed by dye molecules as charge islands. These island molecules could be selectively excited via a scanning tunneling tip in order to induce quantized (and localized) charge transport.

Electron Sources

Figure 6.33 shows a nanostructured integrated electron source, which has been fabricated with the help of three-dimensional electron beam lithography [511]. The electrons are emitted from 4 nm diameter gold nano crystals, which make a tip with a radius of 10 nm and a height of 100 nm. The potential distribution around these micro tips results in a directed electron emission with an emission angle of $\pm 7°$. The total emission current is about $10\,\mu\text{A}$, and the brilliance is about a factor of 10 higher compared to conventional Schottky field-emission sources.

Besides the emitter, the integrated structures from Figure 6.33 also includes a nanoscaled ring electrode, which acts as extractor mesh. Such integrated field emission cathodes, due to their small capacity, could be used in microtriodes, microtubes or for the generation of massive parallel electron beams.

More recently, carbon nanotubes have also shown high potential as field emitter sources, being intrinsically much smaller (diameters in the nanometer range) and much easier and quicker to fabricate.

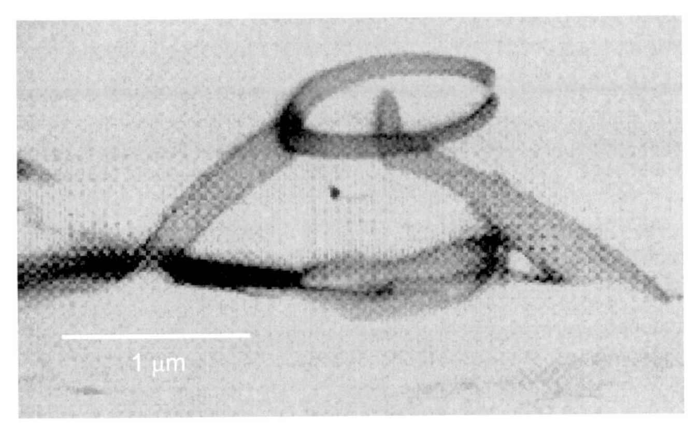

Figure 6.33 Submicron sized field emission electron source. Reprinted with permission from [511]. Copyright 1998 American Vacuum Society.

6.3 Quantum Computers

In the introduction we discussed Moore's law as limit for the possible information density in computer processors. This limit is based on the fact that the nano world obeys quantum mechanical rules which limit the precision with which nanoscaled physical objects can represent the properties of bits, logical operations and so an. Position and momentum of a single atom in a well defined quantum state as a representation of a one-dimensional state cannot be determined with arbitrary precision (Heisenbergs uncertainty relation). Even more important: the measurement process at the atom changes the atom state. However, these limitations might also lead to new possibilities. Such a promising new approach is the quantum computer [5].

In principle it should be possible to obtain and process information directly with the help of quantum mechanics. Instead of using a simple 0/1-system (*bit*) as the basis unit which can exist only in two discrete states, a new basis unit is introduced, called the *qubit*. This unit is prepared as a superposition of discrete states. This way one takes into account the quantum mechanical superposition principle, which introduces, in addition to, the amplitude some phase information. A *qubit* can thus simultaneously store the states '0' and '1' in an arbitrary ratio. This exponentially increases the possibilities of information storage compared to a classical register. A classical 3-bit register can store exactly *one* out of 8 numbers, that is, 000, 001, 010, 011, 100, 101, 110 or 111 (binary 0 to 7). The quantum mechanical analogue can store the 8 numbers *simul-*

taneously in a superposition state, which means that N *qubits* can store 2^N numbers.

The disadvantage of this approach is that the stored information is not simultaneously adressable. The measurement of the state of the quantum register reduces the register to a single value . In order to take advantage of the increased information density one has to perform mathematical operations at the total state (i.e. the superposition of all informations) *before* the system is read out. This could be used in algorithmic search routines for the search on lists which are generated only during the search process, for example for chess programs or for decrypting routines.

The idea of using quantum phenomena for performing calculations [512] and the theoretical basics of 'quantum calculations' are more than twenty years old [513]. Within the last few years, however, one has started to investigate various physical approaches to realize the quantum computer. In the rest of this section we discuss some of the most promising processes.

Ion traps

The overall idea is to trap ions in an electric field, to isolate and cool them and to realize *qubits* via a change of the ion states. All ions together (usually in the form of a linear chain) form the quantum register, within which quantum mechanical calculations are performed. The 'gate'-operations inside the register rely on the entanglement of internal degrees of freedom of the ions (electronic excitation) and the collective movements of all ions in the trap. At the end of the computing process the final state is read via another interaction with laser light.

In a linear 'Paul-trap' [372] ions can be captured and optically cooled so that they form ordered linear structures along the trap axis owing to mutual Coulomb repulsion (Figure 6.34). Keeping the ions in the trap requires a restoring force towards the axis, which is generated by applying a radiofrequency AC voltage between the pole caps, which results in a quadrupole potential with a depth of a few electron volts.

The ions in the trap perform joint ('center-of-mass') or mutual ('breathing') vibrational movements with frequencies ω_n, which can be described as normal modes. In order to generate a well-defined initial vibrational state for the quantum computing, the ions have to be cooled. This is done in general initially via Doppler cooling and subsequently via sideband cooling.

Doppler cooling [514] uses the momentum transfer $\hbar k$ ($k = \lambda_0/2\pi$ denotes the wavevector) following absorption and emission of photons of frequency ω_0 in order to delete kinetic energy of ions. If the ion moves against the laser beam, it can absorb a laser wavelength that is red-shifted

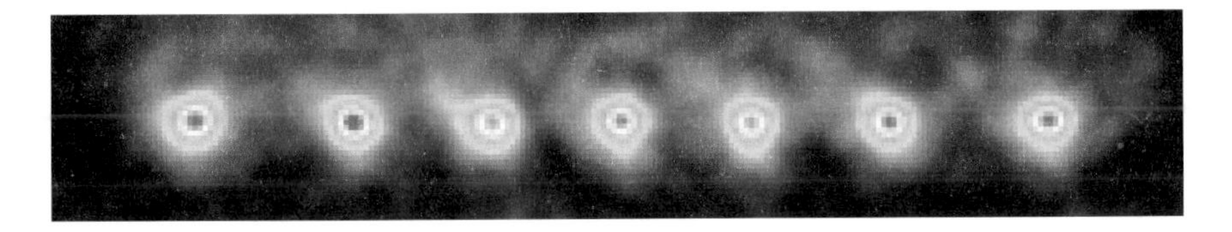

Figure 6.34 Optical image of a linear chain (about 200 μm long) of seven ions, which have been trapped in a Paul trap [373]. Reprinted with permission.

compared to a stationary particle and thus a photon with directed momentum. During such a process the ion is transferred into an excited state. The subsequent emission into the ground state emits a photon with a momentum that has the same magnitude but no correlated direction. Hence the net momenta of the absorbed photons are summed up, while the net momenta of the emitted photons statistically cancel out. As a result, the ion is decelerated by the summed photon momenta (see also Section 5.3). The deceleration per cycle is small (for sodium atoms on the D1-resonance line the decrease in velocity is $0.03\,\mathrm{ms}^{-1}$), making a lot of absorption-emission cycles necessary. The minimum temperature is

$$T_D = \frac{\hbar\gamma}{2k_B} \tag{6.20}$$

and follows from the natural linewidth γ of the excited ionic state. The absolute value is of the order of milli-Kelvin.

At a temperature of a few milli-Kelvin the ion in the trap is not in the ground state with respect to its movements in the trap. This residual vibrational motion affects the absorption spectrum of the ion in the 'dressed state', which can be described as a superposition state between the energetic structure of the free ion and an harmonic oscillator movement in the trap potential: side bands $\omega_0 \pm n\omega_n$ are generated in the absorption spectrum of the ion. If the linewidth connected to the lifetime of the excited ionic state is smaller than the distance between two vibrational states, sideband cooling [515] can be used to transfer the ions into the vibrational ground state in the trap.

In such a case one can use, for example, two laser pulses to induce a Raman transition of the ion into an excited internal state of the red-shifted (low-energetic) side band. The subsequent spontaneous emission into the ground state damps the movement of the ion in the trap by one vibrational quantum. It is important that the recoil energy of the photons is absorbed by the complete trap structure and not by the individual ion alone – hence this laser cooling method is not limited by the recoil energy of the photons[15].

In a trap that is useful for a quantum computer one needs more than one ion in order to fabricate a quantum register from several *qubits*. These ions are coupled mutually via Coulomb forces. The coupling of the vibrational motions makes the cooling process more difficult. Hence, real cooling protocols are in general more complicated compared to the simple process for a single ion as sketched above [517].

Following cooling of the linear chain into the vibrational ground state, it can be used as a quantum register. Selective laser excitation of individual ions stores information into the chain, that is, a quantum mechanical state is stored as a *qubit*. For this purpose the ions have to have a mutual distance of at least a few micrometers in order to be addressed individually within the diffraction-limited focus diameter of the excitation laser. The strong Coulomb repulsion between the ions allows distances of such an order of magnitude (Figure 6.34). The distance between the ions of course also depends on the steepness of the trap potential: the steeper the potential, the smaller the distance between the ions. Sometimes a steep potential is necessary to increase the distance between the energetic states of the ions, which simplifies the cooling process. Fortunately the steepness of the potential is a dynamic parameter, which can be adjusted between filling the trap and setting it into the right state concerning the distance between the ions.

The *qubit* preparation requires an initial clearance of the register via optical pumping into the ionic ground state. The second step involves generation of a quantum mechanical superposition state from electronic ground and excited states via irradiation with a coherent Rabi pulse [16]. The stored quantum information could thus be that the first *qubit* is in the excited state (use of a π pulse) or (for the use of shorter pulses) that it is partially in the excited and partially in the ground state (superposition state).

Following individual adressing of all *qubits*, quantum calculations can be performed via coupling of the *qubits* in the linear chain. The internal state of the individual ion is thus projected onto the total chain vibration (via, e.g. sideband excitation), which in turn influences the state of all other ions. At the end of the calculation the state of the ion regis-

15) The recoil energy forms the limit for the very efficient polarization gradient cooling process [516].

16) Rabi oscillations describe the temporal evolution of the coherently excited system laser plus particle. A complete oscillation corresponds to a circular motion that changes the phase by 2π. The characteristic time scale is provided by the Rabi frequency Ω, which is a measure of the transition probability into the excited state for given laser intensity.

ter is read out via another laser measurement on all individual ions. Of course, at that point the observation, for example, via laser induced fluorescence from appropriate intermediate states has to take into account not only the pure ground- and excited states but also possible mixtures in superposition states.

Experimentally a CNOT operation was realized some time ago [518, 519]. CNOT means 'controlled NOT', that is, the goal-*qubit* will be changed or not, depending on the state of a control-*qubits*. In those experiments, the inner electronic state of the ion as well as the vibrational state in the trap were used as *qubits*.

NMR (Nuclear magnetic resonance) Here, the *qubits* are the nuclear spins of the individual atoms in more complex molecules, while the molecule itself acts as a quantum register. Logic operations are generated via the total angular momentum coupling, while the computational result is read out as a magnetic induction signal. Using NMR a more complex quantum calculation has been performed for the first time [520].

CQED (Cavity quantum electrodynamics) Similar to the ion traps, the *qubit* is the state of the atom prepared by a laser, where the atom is now inserted into a resonator with a very high Q-value. It interacts with a cavity vibration and forms an entangled state. This state can be changed via interaction with additional laser pulses. The realisation of calculational operations is, however, difficult, since a series of individual resonators have to be coupled.

Quantum dots In this case the *qubits* are the elecron spins in the bound quantum dot state. Quantum logic is realized via magnetic field induced coupling of quantum dots.

Optical lattices The atoms in the potential wells of the lattice are the *qubits* (see also Section 5.3). Computations on the register of all atoms are obtained via atom–atom interactions by spatial movement of the atoms depending on their internal states.

Josephson contacts The *qubits* are the quantized charges in small islands, which are coupled depending on their bias voltage via Josephson tunneling. Inductive coupling might allow quantum logics.

Problematic for all these possible realizations of a quantum computer is the loss of coherence during the computational process ('decoherence'). As a result the quantum system is coupled to the surroundings.

This coupling represents a measurement of the state of the system, that is, a projection into the 'classical' world of fixed states. Hence all advantages of quantum computing are lost.

Decoherence occurs for ion traps, for example, via instabilities of the trap potenial, via coupling of the ground state vibration of the linear chain with neighboring vibrational states, via scattering at residual gas particles or via charges at the electrodes of the Paul trap, induced by the ion. This reduces the possible computing time to milliseconds.

In addition to the decoherence restriction, the number of potential problems to be computationally solved with the help of a quantum computer is very limited, while the necessary number of individually adressable *qubits* is orders of magnitude higher than achievable at present. First in 2000 could the entanglement of four atoms be demonstrated [521]. With increasing number of atoms or ions the difficulty for realization grows exponentially. Hence, it is to be expected that the realization of classical computers containing nanoscaled systems (viz., isolated atoms or molecules) will be achieved much earlier than the realization of, real, quantum computers.

6.4 Biology

6.4.1 Characterization of Elementary Units

Elementary units of biology are organic molecules, proteins, enzymes, DNA and supramolecular units such as membranes or vesicles. Important for the insertion into larger complexes, and for an understanding of functionality, is information about mechanical and electric properties of the biomolecules. Hence inter- and intramolecular interactions, structure and electronic behaviour have to be investigated as thoroughly as possible.

An obviously well suited tool for such structural investigations of biological materials is the force microscope [522]. One achieves spatial resolution in the nanometer range, that is, comparable to scanning electron or scanning tunneling microscopy[17] *and* the materials can be investigated in

17) Molecular resolution can be achieved relatively easy under ultrahigh vacuum conditions using the STM. Even under such conditions biological relevant surface processes can be observed, for example, the orientation (chirality) of adsorbed biomolecules, which can be used for biological identification [523,524].

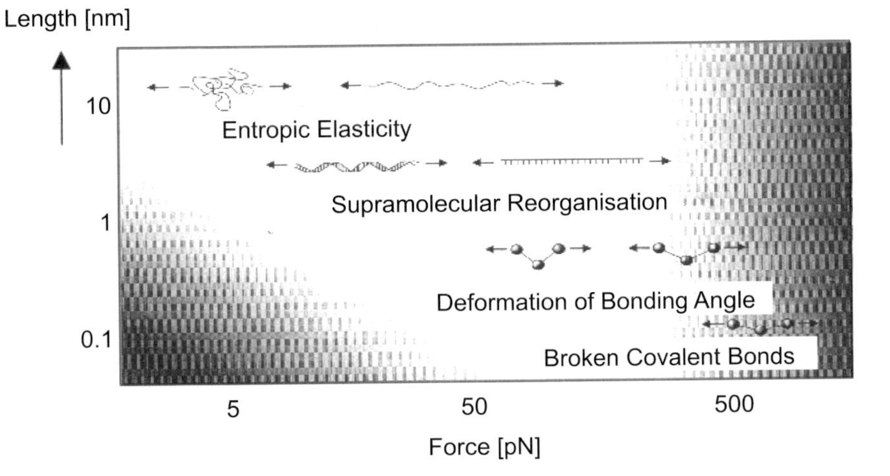

Figure 6.35 Length vs. force scale for biologically relevant units. Reprinted with permission from [526]. Copyright 2000 Elsevier Science Ltd.

their natural environment. Especially problematic are the side views of, for example, nerve cells or structures with large height-width-ratio since the imaging tip has a finite size and it might have been contaminated by biomaterial. In order to improve the imaging, very long and thin tips can be used (e.g. tips that include carbon nanotubes) or one combines AFM and SEM to obtain complementary information [525]. Investigations of living objects *in vivo* are however difficult with AFM since the AFM imaging is a slow method. An alternative in many cases can be confocal microscopy, especially if a resolution in the submicron range is sufficient.

Besides structural investigations AFM also allows one in a unique fashion to investigate mechanical and dynamical properties on a micro-nano scale. Force microscopy on *single* molecules ('force spectroscopy') [526–528] has been shown to be an excellent tool for the investigation of the functionalities of biomolecules in their natural environment (the liquid phase). Using this approach, interaction potentials, protein folding routes and mechanical properties of DNA ('stretch measurements'), molecular motors [529] and selective structural changes of DNA as a result of binding of proteins or drugs [530] could be better understood.

Figure 6.35 shows typical forces and lengths which are of importance for single molecule force spectroscopy. Areas which are not accessible are shaded. In these areas either the covalent bonds are broken (upper right) or the molecular structures become instable (lower left). Within

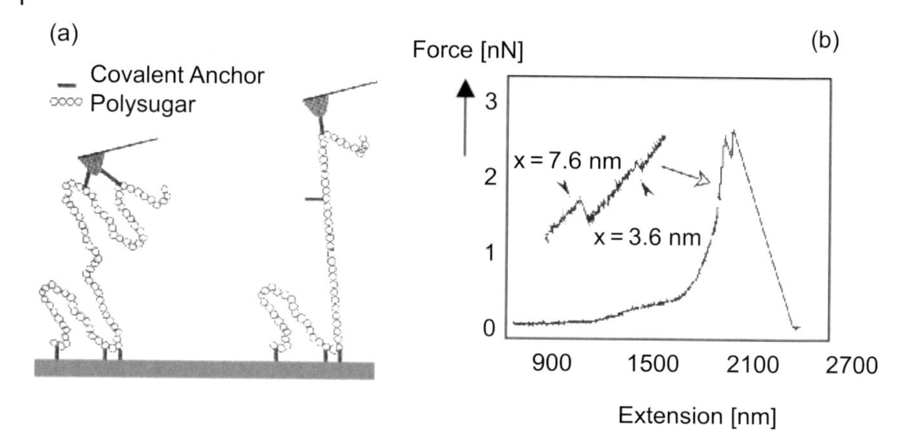

Figure 6.36 (a) Sketch for the binding of an AFM tip to a sugar polymer with subsequent stretching via removing the tip from the surface. (b) Force vs.distance tip-surface. The lower image shows that the stretching of the polymer is accompanied by a break of the bonds to the silicon oxide surface. Reprinted with permission from [531]. Copyright 1999 Science.

the nonshaded areas force spectroscopy can be performed. Figure 6.36a illustrates the idea: the biomolecule to be investigated (a sugar polymer) is covalently bound both to substrate (silicon oxide) and AFM tip (Silicon). If one increases the distance between tip and surface, the molecule unfolds, stretches and eventually the covalent bond between carbon and silicon breaks (Figure 6.36b).

After calibration it is found that the force for scission of the carbon-silicon bond is of the order of 2 nN while a molecule that is bound via a thiole bond to a gold surface detaches after 1.4 nN. A more detailed analysis of the force-distance curve shows a plateau, that is, an increase of molecular length without an increase of applied force. This is found even more expressed in the case of DNA. Double-stranded DNA can be stretched to twice its contour length without breaking. The (reversible) force-induced structural change occurs at 65 pN. Further stretching then leads to a modification of the double helix into two single strands [532].

6.4.2
Nanobionics and Nanobiotechnology

The transition from an understanding of the fundamental biological mechanisms to technological applications leads to 'bionics' [533]. Here one tries to technologically mimic 'natural', evolutionary optimized processes from materials over constructions and motions all the way to system formations.

Especially in microbionics one finds a lot of technologically useful, functionally optimized processes and systems. Historically seen, brittle stars, for example, *Ophiocoma wendtii* (Figure 6.37), were amongst the first living entities which utilized micro collection lenses in order to bundle daylight and so perform photosynthesis even in the entrance areas of caves. In Figure 6.37 their complex form of biological micro lenses is shown, in which a correction of spherical aperture has been achieved via a combination of different radii of curvatures. The whole construction is made from a chalconic skeleton.

In general natural colors result – if they are due to physical effects and not pigments – from interference, diffraction and refraction effects in thin films or periodic arrays of nanoscaled structures [534]. Butterfly wings as photonic materials have been discussed in Section 6.1.3. Butterfly eyes incorporate a biological interference filter which is made from a periodic array of cytoplasm plates with high index of refraction and air gaps with low index of refraction. This interference filter results in a wavelength selective light reflection and thus a metallic appearance of the eyes. The backs of bugs show an intense metallic reflection and the reflected light is often circularly polarized due to imprinted grating structures. Even optical fibers have natural predecessors. In the fibers of each sprouting corn, sunlight is guided from the sprout to the roots in order to direct growth. Recently it has been discovered that deep see sponges (Euplectella) have developed very flexible and rotatable optical fibers which might result in a new kind of technical optical fibers [535].

In the area of very small units, nanobionics is defined as the technological application of functional biological molecules [537][18]. Such artificially modified molecules have been generated naturally, often on the basis of proteins or DNA fragments.

In principle one follows two different approaches. First, the analytical approach, which uses optimized biomolecules as key elements in nanostructures, which have been fabricated via top down (lithographical) techniques. An example are the photodetectors shown in Figure 6.5 with biomolecules as cathodes. Second, the synthetic approach. Here, one tries to generate two- or three-dimensional arrays of meso- or nanoscopic structures with the help of special biomolecules. It has been proven that for such processes, DNA is a very versatile assembler [538]. A good

18) Proteomics, that is, the science to find the key to the folding code of proteins and to understand the relationships between function, sequence and structure of proteins, can also be characterized as a part of nanobionics.

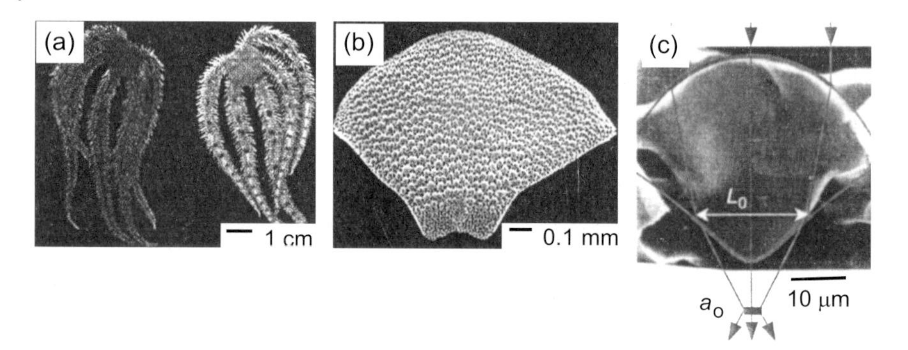

Figure 6.37 Biological micro lenses: (a) *Ophiocoma wendtii* in daylight (left-hand-side) and at night (right-hand-side), showing significant color changes. (b) SEM of a back-plate covered with a collection of micro lenses. (c) SEM of a single micro lens. The drawn line represents a lens with correction for spherical aberration. $a_0 \approx 3$ μm is the focus size, $L_0 \approx 20$ μm is the effective size of the lens aperture. Reprinted with permission from [536]. Copyright 2001 Nature.

example is the generation of nanocrystals with the help of DNA [48]. The formation of an ordered pattern of 2 nm diameter platinum clusters in the pores of a regular matrix of a self-organized protein (the bacterium *Sporosarcina ureae*) has also been reported ('biomolecular templates' [539]).

The basis units which the assembler uses are either molecules with synthetically programmed recognition units or nano particles with well defined surface chemistry. The advantage of biomolecules (peptides, oligonucleotides or proteins) as basis units of course is that molecular recognition units are intrinsically built in.

An example is given in Figure 6.38; namely, a fluorescence microscopy image of a DNA-bridge between two gold electrodes. The gold electrodes have been coated with a layer of oligo-nucleotides in order to enhance sticking of the DNA. The oligo-nucleotides in turn have been anchored on a gold surface with the help of thioles (cf. Section 3.2.2). A silver wire has been fabricated by deposition of silver from solution onto the DNA, initially in the form of clusters and thereafter as a continuous film (Figure 6.38b). The current–voltage curves (Figure 6.39) of the 100 nm diameter silver wire show Ohmic behaviour (linear dependence between current and voltage) or a Coulomb-blockade (no current for slightly negative and positive voltages), depending on the morphological consistence of the wire. Reference measurements without DNA bridge or without silver wire show a resistance larger than $10^{13}\,\Omega$ for

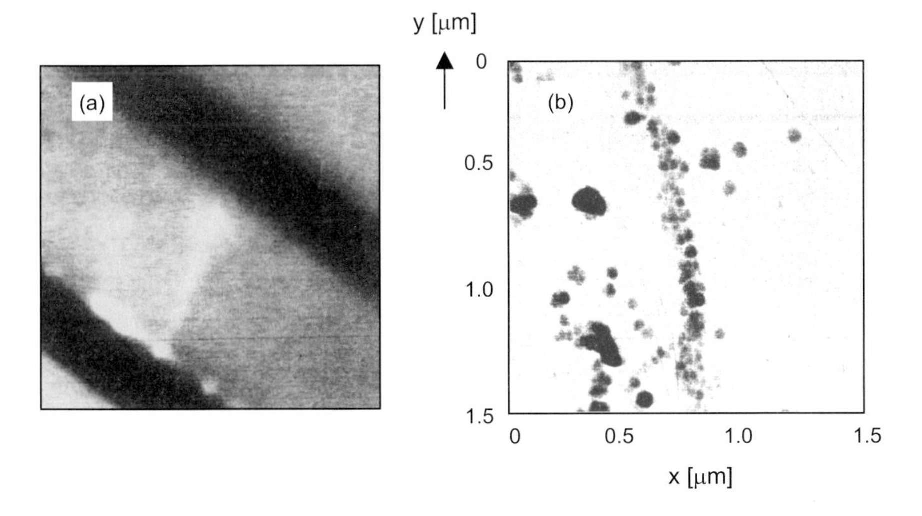

Figure 6.38 (a) Fluorescence image of a DNA bridge between two gold electrodes with a distance of 16 μm. (b) AFM image of a silver wire, generated on the basis of a DNA template between two gold electrodes with a mutual distance of 12 μm. Reprinted with permission from [502]. Copyright 1998 Nature.

a 12 μm wide gap. The DNA used in this example has thus been non-conducting [19].

A fundamental and still unresolved question is whether it is of more advantage for technological applications to apply original biological molecules (e.g. natural DNA) or biomimetic analogues (synthetic oligonucleotides).

6.5 Molecular Carbon Nanostructures

The term molecular nanostructures describes objects which have been grown via bottom up technology from molecular units to a size of a few to a few tens or hundred nanometers. In principle all kinds of clusters (cf. Section 2.2) can be categorized that way, that is, bound assemblies of atoms or molecules. Metallic clusters on solid surfaces, including their electronic dynamics on an ultrafast time scale, are discussed in Section 6.1.4. A good overview over the optical properties of metallic clusters can be found in [445]. See also Section 3.2.2 concerning other

19) Other measurements have shown weak electron transfer through DNA, for example, [540].

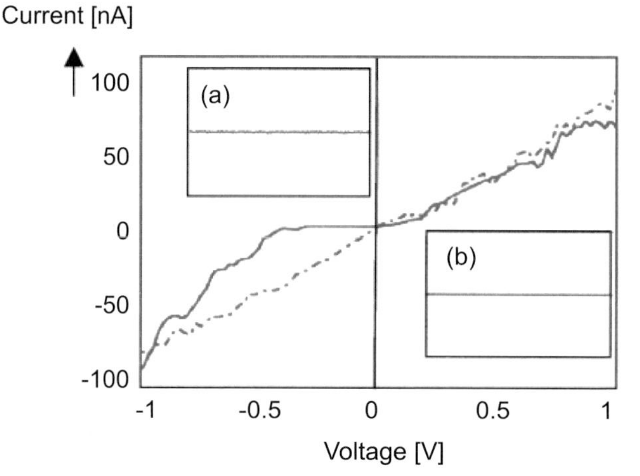

Figure 6.39 I-V-curves for the silver wire shown in Figure 6.38. The inserted images are reference measurements of silver deposition without DNA bridge (a) and DNA bridge without silver deposition (b). Reprinted with permission from [502]. Copyright 1998 Nature.

structures such as epitaxially grown tubes or wires from organic and inorganic molecules.

Carbon Nanostructures

In this section a special class of molecular nanostructures with very high application potential is discussed, namely nanostructures from carbon 'fullerenes' [541–543] and carbon nanotubes. The most famous fullerene molecule is the carbon–60 cluster, C_{60} (Figure 6.40), which has the shape of a soccer ball. As seen immediately, this cluster is characterized by a high symmetry and thus high stability and extraordinary electronic and optical properties, similar to diamond.

In order to generate a curved structure such as the C_{60} cluster from a planar fragment of the hexagonal graphite lattice, topological defects have to be introduced. In addition to the graphite hexagons, pentagons are included. One needs 12 pentagons to generate a curved hexagonal lattice. In all C_{2n} fullerenes one thus finds 12 pentagons and $(n-10)$ hexagons. A stretched fullerene with end caps of 6 pentagons and a large number of hexagons thus forms a 'nanotube' (Figure 6.40c), the diameter of which depends on the diameter of the fullerene which is responsible for the end caps (e.g. 1.2 nm for C_{240}).

The elastic modulus of graphite in its basal plane has an extraordinary high value. It is, thus, obvious that not only a soccerball such as

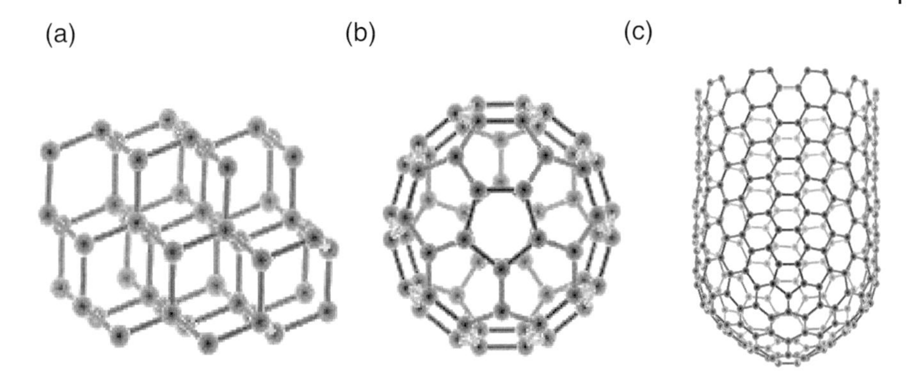

Figure 6.40 Structure of diamond (a), C_{60} (b) and a (10,10) carbon nanotube (c).

the C_{60} but also nanotubes made from graphite possess extraordinary stabilities. Carbon nanotubes are in fact very flexible [544] and stable against stretching (Young's modulus[20] is that of to diamond, approximately 1 TPa [545]).

Within the last years it has been shown that the high hopes one had after the discovery of these materials could only be partially fulfilled. Instead of C_{60}, carbon nanotubes [546] are now showing promise as a molecular nanostructure basis for improved technologies. The main reasons are: carbon nanotubes are dimensionally less restricted compared to clusters; they are mechanically extremely stable; their conductance is variable between semiconducting and conducting; and they can be easily and cheap produced in large amounts. In what follows we discuss C_{60} clusters only briefly with respect to their extraordinary optical nonlinearities and explain in more detail carbon nanotubes.

In Section 5.1 an application of C_{60} molecules in a new kind of polymer solar cells is presented. Ultrathin films from pure C_{60} show very high optical nonlinearities of $\chi^3 \approx 10^{-10}$ esu. This is mainly due to the large number of three-dimensionally delocalized π bonds and an electron gas, which although confined in a symmetric cage with strong boundary conditions, is nearly free inside the cage. If one adsorbs these carbon cages onto a surface, the overall (macroscopic) nonlinearity can even be increased due to multipole interactions. Depolarization effects that are often observed in the case of nonlinearly optically active polymer aggregates are unlikely.

20) Young's modulus Y defines the force that is necessary to bend the tube, $F_b = (\pi^3 Y r^4)(4L^2)$ with L length and r radius of the tube. The spring constant is $k_b = (3\pi Y r^4)/L^3$.

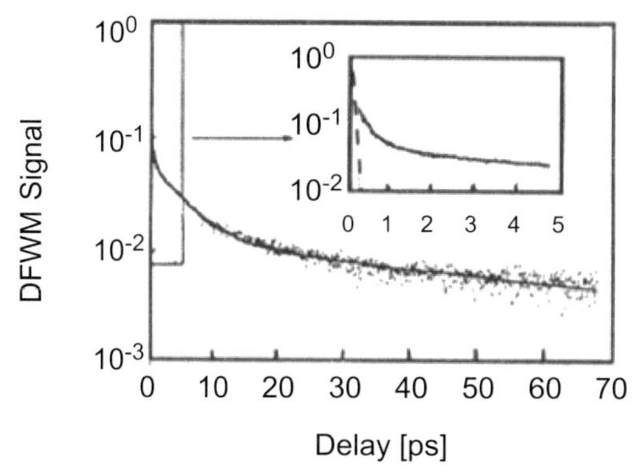

Figure 6.41 DFWM signal as a function of delay time of the backward pump beam at a wavelength of 637 nm [547]. The solid line corresponds to a threefold exponential decay. The dashed line is due to the measured pulse width of the laser (150 fs).

Nonlinear Dynamics in C_{60} Films

One method to measure the nonlinear optical response of a thin C_{60} film on an irradiating light wave (i.e. the hyperpolarizability $\chi^{(3)}$) is to use degenerate four-wave mixing (Figure 4.37). For this the initial laser beam (here a 150 fs laser pulse of wavelength 637 nm, generated by a mode-coupled femtosecond laser) is split into two beams (forward and backward pump), which interfere in the nonlinear medium and form a holographic grating. A third beam (probe beam) is coherently scattered at this grating and generates a phase-conjugate signal beam, the intensity of which is measured using a photomultiplier. As all beams possess the same wavelength (the process is 'degenerate') the wavelength of the signal beam is also predetermined by energy conservation rules. The same is the case with the direction of the emitted signal: momentum conservation forces it to be counter-directed to the probe beam. It can be separated from that beam by a beam splitter.

The signal intensity is proportional to the product of the powers of the three irradiating lasers, the square modulus of the nonlinear susceptibility of third order and the square of the interaction length, which in the case of a 10 nm thick C_{60} film is small. However, the large nonlinearity of the films allows one to measure time-dependent signal intensities applying only relatively small irradiances of the order of gigawatts per square centimeter. After adjusting the probe beam onto the optimum temporal overlap of all three partial beams, one delays the backward pump beam with respect to the forward pump beam and obtains as a function of this delay the signal intensity that is plotted in Figure 6.41.

Within the pulse width of the laser the signal intensity decreases strongly. Hence, it is probable that this part of the signal is generated by a coherent polarization grating of the π-electrons, which exists as long as the laser beams irradiate the material. In the following, the signal decreases with two further exponential time constants of a few hundred femtoseconds and a few picoseconds, respectively. The origin of these signals are gratings of excited electrons in the C_{60} film. The fast component corresponds to the direct decay of the population of the first excited singlet state S_1, while the slow component results from the decay of the long-living triplet state T_1, which has been generated from the singlet state via configuration interaction ('intersystem crossing'). By use of an appropriate laser wavelength the contribution of these long-living components and thus the optical response time of this thin film switch can be varied between femto- and picoseconds.

Nanotubes

Single wall nanotubes (SWNT) possess a diameter of about 1 to 1.5 nm, depending on the size of the half-fullerene which constitutes the end caps, and thus depending on the fabrication method[21]. Multiple wall nanotubes(MWNT) are similar to hollow graphite fibers and consist of concentric cylinders around a central hole. The distance between individual graphite layers is 0.34 nm.

A simple production method for nanotubes is CVD ('chemical vapor deposition'). Here, a precursor (e.g. iron nitrate, $Fe(NO_3)_3$) is adsorbed on the substrate to be coated with nanotubes and heated to about 1000 K. Metal clusters are generated, which serve as nucleation sites for the following growth in an acetylene (C_2H_2) atmosphere.

Depending on their internal structure the tubes can be metallic or semiconducting and therefore serve directly as nanowires or metal/semiconductor connects. Along their preferred directions the conductivity can be of the order of $10^5\,\mathrm{m}^{-1}$; however, for MWNTs this value depends strongly on structural defects or elastic deformations. In SWNTs, due to their high symmetry the molecular wavefunctions are delocalized over the whole tube length. The SNWTs behave like coherent quantum wires [550].

Owing to the many possibilities to roll carbon layers to tubes, one finds tubes with different helicities (Figure 6.42c). Usually nanotubes of different helicity are characterized by the indices (n_1, n_2), which represent the multiplicity of the observed lattice vectors $R = n_1 a_1 + n_2 a_2$ with respect to the primitive lattice vectors (a_1, a_2) of the carbon layer. Tubes of

21) Production via laser ablation of carbon, for example, results in a diameter of 1.38 nm [548].

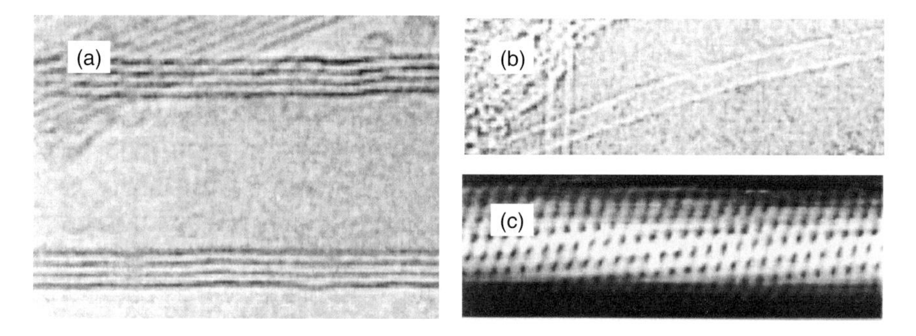

Figure 6.42 TEM images of (a) typical multiwall nanotubes (MWNTs) and (b) a single wall nanotube (SWNT) Image (c) is an STM image of a semiconducting SWNT, clearly showing a helical structure. The diameter of the SWNT is about 1.2 nm. Reprinted with permission from [549]. Copyright 1999 American Chemical Society.

type $(n,0)$ are called 'zigzag' tubes, such of type (n,n) 'armchair'. Calculations predict that the (n,n) nanotubes are metallic, whereas all other nanotubes are either metallic (e.g. (3,0),(4,1),(5,2) ...) or semiconducting (e.g. (1,0),(2,1),(3,2) ...).

Conductivity and stability of nanotubes can be used to improve mechanical stability and electric conductivity [22]of conjugated luminescent polymers by fabricating a composite material [551]. The electronic characteristics of organic light-emitting diodes (OLEDs) can also be improved with the help of nanostructures in the form of nanotubes. This concerns, for example, the equilibrium between injection rates of electrons and holes; the transport of charge carriers as polarons; the recombination rate to singlet-excitons; or the radiative decay of the excitons.

Further interesting applications of carbon nanotubes are as single molecule field effect transistors [552] or as efficient field emission electron sources. If one wants to obtain high current densities with field emission electron sources, a normal problem is that the use of sharp tips (which optimize the field enhancement) leads to a strong decrease of the emission area. A possible solution is to use a matrix of micrometer sized, narrowly packed emission tips. Carbon nanotubes with field enhancement factors of 1000 [553] can be fabricated in large amounts for such a purpose and they can be packed to densities of $10^6\,\mathrm{cm}^{-2}$ via microcontact printing (Section 3.1.2) [554]. The possible total emission current of such matrices depends on the distance between the tubes. Measurements and calculations suggest that optimum emission is achieved for a

22) For example, the conductivity of a specific plain polymer was $2 \times 10^{-10}\,\mathrm{Sm}^{-1}$, that of the filled polymer $3\,\mathrm{Sm}^{-1}$.

mutual distance of twice the height of the tubes [555]. For straight tubes with 1 μm height this results in a density of $2.5 \times 10^7\,\mathrm{cm}^{-2}$. For an emission current between 0.1 and 1 μA per tube, total emissions of up to the Ampere range per centimeter squared can be obtained.

6.6
Micro- and Nanomechanics

Movements on the nanometer scale are dominated by different compared to the corresponding macroscopic movements. Thus, nanomechanics is not simply down-scaled classical mechanics. Friction, for example, dominates since the volume-to-surface ratio in nanometric dimensions is shifted towards the surface. The microscopic mechanisms of friction, in turn, are not completely understood, thus requiring quantitative measurements of dynamical phenomena on a nanometer-scale.

An important tool for this 'nanotribology' [556] is the force microscope, which allows one to measure the lateral forces induced by the friction between tip and surface. Quantitative nanotribology becomes possible via optimization of the tip for lateral movements and minimization of the sensitivity to vertical and torsional movements [557].

Once the mechanical properties are understood on a nanoscale, nanoscaled electronic and mechanical components have to be combined in order to produce functionable NEMS (nano-electromechanical systems), the miniaturized variants of MEMS (micro-electromechanical systems)[558,559] and BioMEMS.

Traditional MEMS are produced with the help of top-down technology (lithographically) on the basis of semiconductor materials, usually silicon. MEMS are 'smart' three-dimensional chips, which combine sensors and actuators for active response at varying environmental parameters. Typical dimensions are in the range of micrometers. Faster (GHz resonance frequencies, that is, pico- or femtosecond time resolution), lighter (masses of femtograms) and more sensitive (force sensitivity a few atto-Newtons, mass sensitivity a few molecules) three-dimensional optoelectromechanical components should be achievable via scaling into the submicron range (NEMS) [560]. In this range the limits of conventional lithography are easily reached. However, new methods such as nanoimprint or 'soft lithography' should allow one to produce MEMS and NEMS from 'soft' materials. Materials of this kind are, for example, elastomers such as polydimethylsiloxane, from which microminiaturized pumps and valves can be fabricated [561]. Bottom-up methods are also applicable to NEMS fabrication.

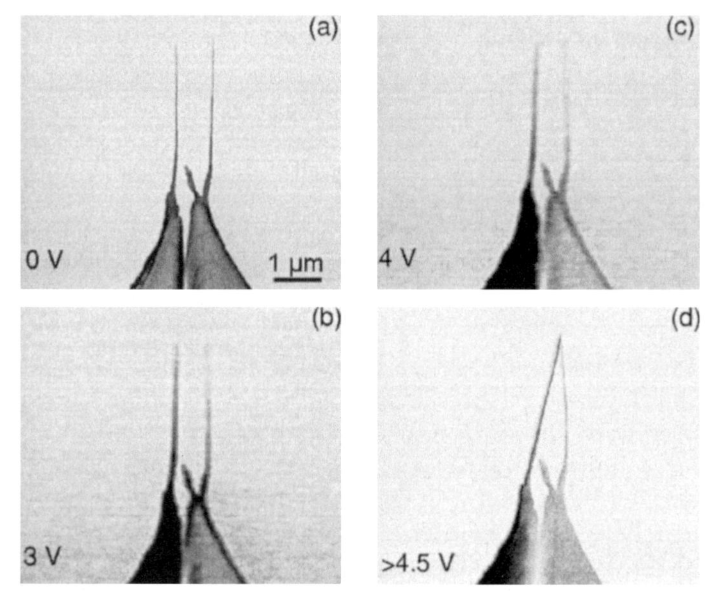

Figure 6.43 'Nano tweezer': Carbon nanotubes at a force microscope tip. As a function of applied voltage a movement of the tweezer is seen in the SEM. Reprinted with permission from [563]. Copyright 2001 American Institute of Physics.

In the following a few examples of nanoscaled mechanical components are discussed, examples which could find application in such NEMS.

Tweezers

Due to their extraordinary mechanical and electric properties carbon nanotubes can be used as 'nano-tweezers' with diameters of a few ten nanometers (in the case of MWNTs) and length of a few micrometers. This has been demonstrated for nanotubes on glass rods [562] and also for nanotubes on the tip of a force microscope [563] (Figure 6.43).

In order to fabricate the AFM based nanotweezer, the silicon tip of the AFM was covered with a thin Ti/Pt film, which had been cut into two using an ion beam. The resulting pieces were contacted separately with aluminum leads. Subsequently nanotubes were attached within a SEM to the separated pieces of the AFM tip and were glued with the help of a carbon film. If one applies a voltage of a few Volts between the tubes, they attract each other electrostatically. The resulting force acts against the strain of the tubes and is stronger than this retracting force for potential differences larger than 4.5 V (Figure 6.43). Since the force is elastic, the resulting tweezer movement is reversible. However, due to the van der Waals interaction between the nanotubes an energy minimum also exists for closed tweezers – therefore a slightly larger voltage is necessary to open the tweezers again.

Although the absolute value is only a few ten nano-Newton (nN), the force exerted by the tweezers is large enough to move nanoscaled objects against the gravitational force or against weak electrostatic and van der Waals forces [564] [23]. This has been demonstrated for 500 nm diameter polystyrole clusters in [562].

Another advantage of carbon nanotubes as nano tweezers is their conductivity, which allows one to fabricate a conducting connection to the nano world. As a result the electric tunneling of individual, nanoscaled objects can be investigated. Ideally the Greens function of a single electron between two local tunnel connections should become measurable [565]. This is one of the most detailed pieces of information one can acquire about the local electronic material properties.

Disadvantageous for the use of nanotubes as nanotweezers, for example, compared to the use of light forces in 'light tweezers' (Section 5.3) is that a voltage of a few volts has to be applied between the tips. For a tip distance of a few hundred nanometers this corresponds to an electric field of a few ten million volts per meter. Organic molecules or biological systems are heavily influenced or even destroyed by such strong fields. Especially for nanobiological applications, other nanotweezers have been developed which work without electric field between the manipulating tips (Figure 6.44 [566]).

The basic idea is to take advantage of voltage differences and thus electrostatic forces between one of the tweezer arms and a supporting arm – consequently there is no electric field within the manipulating arms of the nanotweezers. In this case conventional UV lithography and reactive ion etching are used to generate a silicon oxide structure made of four arms. This structure is subsequently coated via electron beam thin film deposition with layers of 10 nm Ti and 80 nm Au, an so the structure becomes conducting. At the tips of the inner arms the carbon nanotweezers are fabricated via focussing of an electron beam in the SEM and following dissociation of hydrocarbon molecules from a very dilute background gas. This fabrication method allows simultaneous observation and manipulation of nanotweezers and thus a large flexibility in fabrication.

23) The force actuated by the nano tweezer is large compared to typical biophysical forces. E.g., the force of the myosin/aktin motor in muscle tissue is of the order of 1 pN, and membranes can be deformed with a few tenth of a pico-Newton.

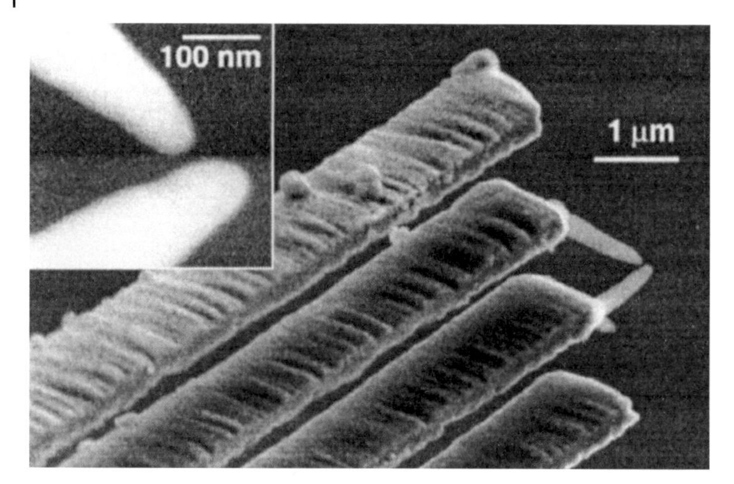

Figure 6.44 SEM images of carbon nanotweezers, which have been mounted on metal-coated silicon oxide arms. The tweezer arms are moved with the help of a voltage difference between the outer arms. Typical dimensions are: distance between the arms 750 nm, diameter of the tweezers at the basis 180 nm, distance between the tips 25 nm. Reprinted with permission from [566]. Copyright 2001 Institute of Physics.

Until now these tweezers have not been used for the mechanical manipulation of nanoscaled biological objects. Measurements have, however, shown that forces up to 10^{-5} N are possible before the tips break. In general, the carbon tips deform elastically to a large extent before they break. Possible future developments are the fabrication of conducting nanotweezers by coating them with thin Ti/Au films and a biocompatible version, which is coated with a salt water resistant nitride film. In this combination a directed manipulation of individual ion channels in cell membranes should become possible.

Rotating Systems

Figure 6.45a is an example for a complex micromechanical system: two microstructured cogwheels can rotate around an axis mounted onto a glass plate. The cogwheels are connected to a rotor (dotted arrow), which is engaged by light forces (Section 5.3) (Figure 6.45b). Both the firmly installed cogwheels and the rotor have been fabricated via two-photon photopolymerisation. The method allows a spatial resolution of 500 nm for production, hence three-dimensional elements with characteristic sizes of a few micrometer can be fabricated (see also Section 3.1.1). Another example for a functioning micromachine (a micro oscillator) is shown in Figure 3.6. This machine has been fabricated with a resolution of 150 nm also using two-photon photopolymerisation.

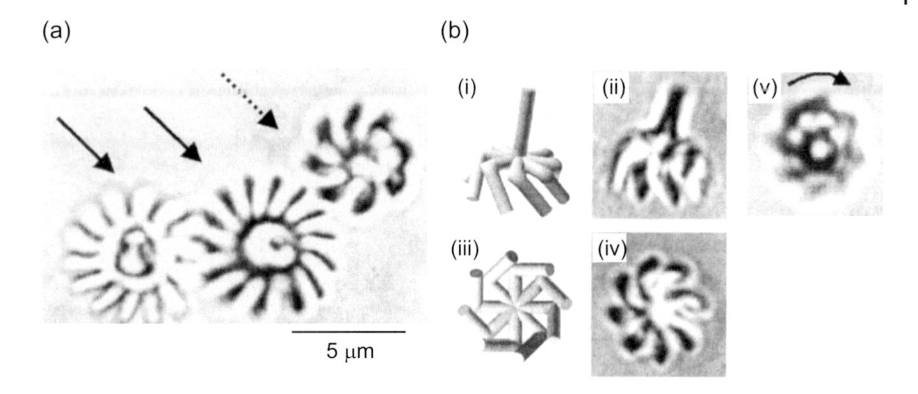

Figure 6.45 (a) SEM image of a micro machine made from two cogwheels with fixed axes (solid arrows), which are actuated by a rotor (dotted arrow). (b) Details of the light-driven rotor. In (i) and (ii) the rotor is not mounted, in (iii) and (iv) it is stabilized in the laser focus. Image (v) is a snapshot during the rotational movement. Reprinted with permission from [77]. Copyright 2001 American Institute of Physics.

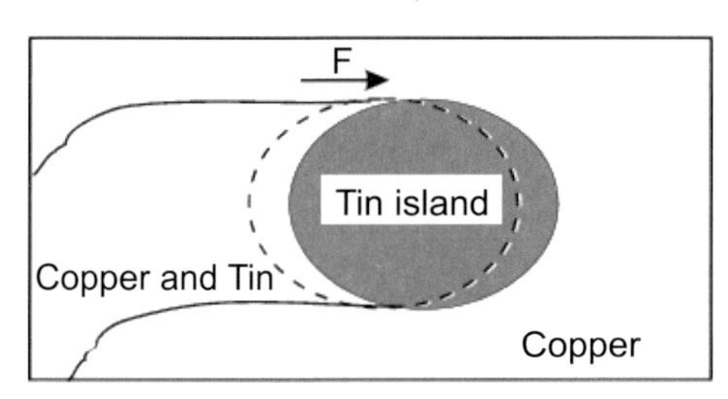

Figure 6.46 Force on a two-dimensional Tin island on top of a Cu(111) surface. The change of surface free energy results in a force for movement along the surface, analogous to the movement of particles at liquid surfaces. The latter phenomenon has been described as early as the 17th century for Kampfer particles on water surfaces.

Actuators

The above section discussed examples of nanoscaled mechanical systems that utilize light as the driving force. In principle, the necessary light can be guided with significant intensity via micrometer-scaled waveguides to the miniaturized elements and can be redistributed in the optical near field. Hence, the dimension of the light source is not necessarily a restriction for the possible packing density. However, it would be much more attractive if the chemical energy could be converted directly into work on the nanometer scale. An open question is, which sources such nanomotors could be [567] and how would they look?

One possibility is to use the resulting chemical energy from an exothermic surface reaction. This has been demonstrated for the example of 100 nm diameter, two-dimensional Zn islands, which migrate along a (111) Cu surface via exchange of Zn with Cu atoms [568]. Figure 6.46 shows schematically the driving principle. At the interface between the Zn islands and Cu an exothermic exchange between Zn and Cu atoms occurs. This exchange becomes energetically more favorable for moving islands since fresh Cu comes into contact with Zn. The direction of movement is given by the fact that the Zn island is repelled from the mixed Cu/Zn phase.

The island hence moves forward by looking for fresh Cu surfaces, which results in a complex two-dimensional movement although the energetics of the driving mechanism is very simple. The power of this chemical motor is significant [567] and comparable, as power per mass, with that of a modern car motor, that is, a few tenths of a HP per kilogram. By structuring the surface and adding new atoms from the gas phase to compensate the removal of atoms, an even more directed application of the principle of such nanomotors should become possible.

In supramolecular chemistry a number of rather developed alternative motors on a molecular scale can be found [529], namely 'molecular actuators', that is, molecules which change form as a response to external forces (e.g. light) and thus in principle do mechanical work. Examples for systems, which can be switched reversibly between two states, are *trans-cis* isomers[24] as a result of light irradiation or rotaxanes [569] (ring-shaped molecules, which can be moved along linear molecules) and catenanes [570] (mutually twisted rings). Motors on a molecular basis have to be somewhat further advanced in order to irreversibly change, for example, chemical into mechanical energy[25]. Artificially tailored molecules allow one to achieve this, as has been demonstrated with the help of light-driven, directed rotation of an alkene along a carbon double bonding [574] as well as a rotation, induced via a thermally induced isomerisation reaction [575]. The molecular details are complex, and the generation of such molecular motors from less than 100 atoms takes a lot of time. The 'bottom-up' character of the method, however, allows one to produce simultaneously a few 10^{19} motors – definitely more than the industrial revolution has generated on a macroscopic level so far.

24) *Trans* isomers have an elongated form, *cis* isomers a bent form.

25) The F_0F_1-ATP synthase in biology shows as a rotating motor for synthesis of ATP how to achieve such a process [571]. Natural linear motors are enzymes such as myosin or kinesin, which use the energy gained from ATP hydrolysis to move along muscle filaments [572]. Recently the movement of a biological motor has been directed via microtubules, which have been steered along kinesin rails [573].

Problems

Problem 6.1 What is the main difference between a photonic and an electronic crystal? Describe characteristic dimensions. How can a photonic bandgap help to tailor the optical appearance of objects?

Problem 6.2 Take into account the Schrödinger equation for matter waves. What is the optical analogue for photons?

Problem 6.3 What are main differences between photons and electrons? Consider vector character, spin and charges.

Problem 6.4 Is there in general a quantum mechanical reason for size-dependent optical properties of plasmonic structures? What other physical effects play an important role?

Problem 6.5 Consider a nanorod-shaped metallic nanoparticle. How many plasmon resonances do you expect?

Problem 6.6 Future nanoelectronics needs fast switches. Compare the minimum switching time of molecular electronics, based on, for example, rearrangement of molecules and crystal nanoelectronics, based on, for example, nonlinear scattering phenomena.

Problem 6.7 Self assembled monolayers can be used as dielectric spacer layers of very high quality. What is the main reason why they anyway cannot be used as dielectric medium for, for example, nanocapacitors of more than a few ten nanometers size?

Problem 6.8 The wings of some butterflies show wide angle and bright color reflection. Why is the responsible underlying structure of the wings sometimes called a photonic band gap structure?

Problem 6.9 What is the main obstacle for the future application of carbon nanotubes in nanoelectronics?

Problem 6.10 The penetration depth of light into matter depends on the absorptivity. Therefore dielectric materials cannot be investigated optically with high surface sensitivity. Why is this different for semiconductors and metals?

Problem 6.11 The surface sensitivity of nonlinear optics can be of the order of a nanometer. What determines this size scale?

References

1 R.P. Feynman. There's plenty of room at the bottom. *Engineering and Science*, 23:22–36, 1960.

2 G.E.Moore. Cramming more components onto integrated circuits. *Electronics*, 38(8):114–117, 1965.

3 G.E.Moore. *Intel Developer Update Magazine*, 2, 1997.

4 SIA. International technology roadmap for semiconductors, 2000.

5 D. Bouwmeester, A.K. Ekert, and A. Zeilinger, editors. *The Physics of Quantum Information: Quantum Cryptography, Quantum Teleportation, Quantum Computing*. Springer, Berlin, 2000.

6 R.A. Freitas. *Nanomedicine, Vol.1: Basic Capabilities*. Landes Bioscience, Georgetown, 1999.

7 K. Drexler. *Nanosystems: Molecular Machinery, Manufacturing, and Computation*. John Wiley, New York, 1992.

8 M.Schiek, F.Balzer, J.Brewer, K.Al-Shamery, and H.-G. Rubahn. Organic molecular nanotechnology. *SMALL*, 4:176, 2008.

9 R. Tredgold. *Order in Thin Organic Films*. Cambridge University Press, Cambridge, 1994.

10 R. Schinke. *Photodissociation Dynamics: Spectroscopy and Fragmentation of Small Polyatomic Molecules*. Cambridge Monographs on Atomic, Molecular, and Chemical Physics. Cambridge University Press, Cambridge, 2006.

11 H.-G. Boyen, G. Kästle, F. Weigl, P. Ziemann, G. Schmid, M.G. Garnier, and P. Oelhafen. Chemically induced metal-to-insulator transition in Au_{55} clusters: Effect of stabilizing ligands on the electronic properties of nanoparticles. *Phys. Rev. Lett.*, 87:276401, 2001.

12 H. Göhlich, T. Lange, T. Bergmann, and T.P. Martin. Electronic shell structure in large metallic clusters. *Phys. Rev. Lett.*, 65:748–751, 1990.

13 G. Rubio-Bollinger, S.R. Bahn, N. Agraït, K.W. Jacobsen, and S. Vieira. Mechanical properties and formation mechanisms of a wire of single gold atoms. *Phys. Rev. Lett.*, 87:026101, 2001.

14 P.M. Petroff, A. Lorke, and A. Imamoglu. Epitaxially self-assembled quantum dots. *Phys. Today*, 54(5):46–52, 2001.

15 F. Frankel and G.M. Whitesides. *On the Surface of Things: Images of the Extraordinary in Science*. Harvard University Press, Harvard, 2008.

16 V.I. Klimov and M.G. Bawendi. Ultrafast carrier dynamics, optical amplification, and lasing in nanocrystal quantum dots. *MRS Bulletin*, 26(12):998, 2001.

Basics of Nanotechnology: 3rd Edition. Horst-Günter Rubahn

ISBN: 978-3-527-40800-9

17 V.I. Klimov, A.A. Mikhailovsky, S. Xu, A. Malko, J.A. Hollingsworth, C.A. Leatherdale, and and M.G. Bawendi H.-J. Eisler. Optical gain and stimulated emission in nanocrystal quantum dots. *Science*, 290:314–317, 2000.

18 R.G. Neuhauser, K.T. Shimizu, W.K. Woo, S.A. Empedocles, and M.G. Bawendi. Correlation between fluorescence intermittency and spectral diffusion in single semiconductor quantum dots. *Phys. Rev. Lett.*, 85:3301–3304, 2000.

19 M. Kuno, D.P. Fromm, H.F. Hamann, A. Gallagher, and D.J. Nesbitt. Nonexponential "blinking" kinetics of single CdSe quantum dots: A universal power law behavior. *J. Chem. Phys.*, 112:3117, 2000.

20 X. Peng, L. Manna, W. Yang, J. Wickham, E. Scher, A. Kadavanich, and A.P. Alivisatos. Shape control of CdSe nanocrystals. *Nature*, 404:59–61, 2000.

21 C.R.C. Wang, S. Pollack, D. Cameron, and M.M. Kappes. Optical absorption spectroscopy of sodium clusters as measured by collinear molecular beam photodepletion. *J. Chem. Phys.*, 93:3787, 1990.

22 Y. Wang, C. Lewenkopf, D. Tomanek, and G. Bertsch. Collective electronic excitations and their damping in small alkali clusters. *Chem. Phys. Lett.*, 205:521–528, 1993.

23 Ch. Ellert, M. Schmidt, Th. Reiners, and H. Haberland. Transition of the electronic response from molecular-like to jellium-like in cold, small sodium clusters. *Z. Phys. D*, 39:317–323, 1997.

24 Th. Reiners, Ch. Ellert, M. Schmidt, and H. Haberland. Size dependence of the optical response of spherical sodium clusters. *Phys. Rev. Lett.*, 74:1558–1561, 1995.

25 K. Selby, M. Vollmer, J. Masiu, V. Kresin, W.A. De Heer, and W.D. Knight. Surface plasma resonance in free metal clusters. *Phys. Rev. B*, 40:5417–5427, 1989.

26 H. Hövel, S. Fritz, A. Hilger, U. Kreibig, and M. Vollmer. Width of cluster plasmon resonances: Bulk dielectric functions and chemical interface damping. *Phys. Rev. B*, 48:18178–18188, 1993.

27 M. Scharte, R. Porath, T. Ohms, M. Aeschlimann, J.R. Krenn, H. Ditlbacher, F.R. Aussenegg, and A. Liebsch. Do Mie plasmons have a longer lifetime on resonance than off resonance? *Appl. Phys. B*, 73:305–310, 2001.

28 P. Apell and D. Penn. Optical properties of small metal spheres: Surface effects. *Phys. Rev. Lett.*, 50:1316–1319, 1983.

29 A. Liebsch. Surface-plasmon dispersion and size dependence of mie resonance: Silver versus simple metals. *Phys. Rev. B*, 48:11317–11328, 1993.

30 A. Liebsch. *Electronic Excitations at Metal Surfaces*. Plenum Press, New York, 1997.

31 C. Yannouleas. Microscopic description of the surface dipole plasmon in large Na_N clusters. *Phys. Rev. B*, 58:6748–6751, 1998.

32 A. Broers. In R. Bakish, editor, *First International Conference on Electron and Ion Beam Science and Technology*, New York, 1964. Wiley.

33 G.M. Wallraff and W.D. Hinsberg. Lithographic imaging techniques for the formation of nanoscopic features. *Chem. Rev.*, 99:1801–1822, 1999.

34 T. Bloomstein, M. Horn, M. Rothschild, R. Kunz, S. Palmacci, and R. Goodman. Lithography with 157 nm lasers. *J. Vac. Sci. Techn. B*, 15:2112–2116, 1997.

35 D.J. Elliott. *Ultraviolet Laser Technology and Applications*. Academic Press, New York, 1995.

36 M. Peckerar, F. Perkins, E. Dobisz, and O. Glembocki. Issues in nanolithography for quantum effect device manufacture. In P. Rai-Choudhury, editor, *Handbook of Microlithography, Micromachining and Microfabrication*, pages 681–763, London, 1997. IEEE Materials and Devices Series.

37 M. Levenson, N. Viswanathan, and R. Simpson. Improving resolution in photolithography with a phase-shifting mask. *IEEE Trans. Electr. Dev.*, ED-29:1828–1836, 1982.

38 K. Ronse, M. Op de Beeck, L. Van den hove, and J. Engelen. Fundamental principles of phase shifting masks by Fourier optics: Theory and experimental verification. *J. Vac. Sci. Technol. B*, 12:589–600, 1994.

39 D. Elliott. Pathway to the gigabit memory. *Laser and Optronics*, June/July 1993.

40 K. Ronse, R. Pforr, R. Jonckheere, and L. Van den hove. Attenuated phase shifting masks in combination with off-axis illumination: Towards quarter-micron DUV lithography for random logic applications. *Microel. Eng.*, 23:133–138, 1994.

41 P. Rai-Choudhury, editor. *Handbook of Microlithography, Micromachining and Microfabrication*, London, 1997. IEEE Materials and Devices Series.

42 E. Kratschmer. Verification of a proximity effect correction program in electron-beam lithography. *J. Vac. Sci. Technol.*, 19:1264–1268, 1981.

43 G. Owen and P. Rissman. Proximity effect correction for electron beam lithography by equalization of background dose. *J. Appl. Phys.*, 54:3573–3581, 1983.

44 M.A. Gesley and M.A. McCord. 100 kV ghost electron beam proximity correction on tungsten x-ray masks. *J. Vac. Sci. Technol. B*, 12:3478–3482, 1994.

45 NN. *Optics and Lasers Europe*, page 21, April 2001.

46 S.D. Berger, J.M. Gibson, R.M. Camarda, R.C. Farrow, H.A. Huggins, J.S. Kraus, and J.A. Liddle. Projection electron-beam lithography: A new approach. *J. Vac. Sci. Techn. B*, 9:2996–2999, 1991.

47 Y. Wang and N. Herron. Nanometer-sized semiconductor clusters: Materials synthesis, quantum size effects, and photophysical properties. *J. Phys. Chem.*, 95:525–532, 1991.

48 A.P. Alivisatos. Perspectives on the physical chemistry of semiconductor nanocrystals. *J. Phys. Chem.*, 100:13226–13239, 1996.

49 J. Shah. *Ultrafast Spectroscopy of Semiconductors and Semiconductor Nanostructures*. Springer-Verlag, Berlin, New York, 1996.

50 J.H. Fendler, editor. *Nanoparticles and Nanostructured Films*, Weinheim, 1998. Wiley-VCH.

51 H.W. Deckman, J.H. Dunsmuir, S. Garoff, J.A. McHenry, and D.G. Peiffer. Macromolecular self-organized assemblies. *J. Vac. Sci. Technol. B*, 6:333–336, 1988.

52 J.C. Hulteen and R.P. Van Duyne. Nanosphere lithography: A materials general fabrication process for periodic particle array surfaces. *J. Vac. Sci. Technol. A*, 13:1553–1558, 1995.

53 F. Burmeister, J. Boneberg, and P. Leiderer. Mit Kapillarkräften zu Nanostrukturen. *Physik. Bl.*, 56(4):49–51, 2000.

54 J. Boneberg, F. Burmeister, C. Schäfle, P. Leiderer, D. Reim, A. Fery, and S. Herminghaus. The formation of nano-dot and nano-ring structures in colloidal monolayer lithography. *Langmuir*, 13:7080–7084, 1997.

55 G. Timp, R.E. Behringer, D.M. Tennant, J.E. Cunningham, M. Prentiss, and K.K. Berggren. Using light as a lens for submicron, neutral-atom lithography. *Phys. Rev. Lett.*, 69:1636–1639, 1992.

56 W.R. Anderson, C.C. Bradley, J.J. McClelland, and R.J. Celotta. Minimizing feature width in atom optically fabricated chromium nanostructures. *Phys. Rev. A*, 59:2476–2486, 1999.

57 E.F. Wassermann, M. Thielen, S. Kirsch, A. Pollmann, H. Weinforth, and A. Carl. Fabrication of large scale periodic magnetic nanostructures. *J. Appl. Phys.*, 83:1753, 1998.

58 D.Haubrich, D.Meschede, T.Pfau, and J.Mlynek. *Physik.Blaetter*, 53:523, 1997.

59 K.K. Berggren, A. Bard, J.L. Wilbur, J.D. Gillaspy, A.G. Helg, J.J. McClelland, S.L. Rolston, W.D. Phillips, M. Prentiss, and G.M Whitesides. Microlithography by using neutral metastable atoms and self-assembled monolayers. *Science*, 269:1255–1257, 1995.

60 F. Lison, H.-J. Adams, D. Haubrich, M. Kreis, S. Nowak, and D. Meschede. Nanoscale atomic lithography with a cesium atom beam. *Appl. Phys. B*, 65:419–421, 1997.

61 S.B. Hill, C.A. Haich, F.B. Dunning, G.K. Walters, J.J. McClelland, R.J. Celotta, and H.G. Craighead. Patterning of hydrogen-passivated Si(100) using $Ar(^3P_{0,2})$ metastable atoms. *Appl. Phys. Lett.*, 74:2239, 1999.

62 W.W. Duley. *UV Lasers: Effects and Applications in Materials Science*. Cambridge University Press, Cambridge, 2005.

63 S. Preuss and M. Stuke. Subpicosecond ultraviolet laser ablation of diamond: Nonlinear properties at 248 nm and time-resolved characterization of ablation dynamics. *Appl. Phys. Lett.*, 67:338–340, 1995.

64 H. Varel, D. Ashkenasi, A. Rosenfeld, M. Wähmer, and E. Campbell. Micromachining of quartz with ultrashort laser pulses. *Appl. Phys. A*, 65:367–373, 1997.

65 D. Ashkenasi, H. Varel, A. Rosenfeld, S. Henz, J. Herrmann, and E. Campbell. Application of self-focusing of ps laser pulses for three-dimensional microstructuring of transparent materials. *Appl. Phys. Lett.*, 72:1442–1444, 1998.

66 H.K.Toenshoff et al. *LaserOpto*, 31(3):68, 1999.

67 J.H. Klein-Wiele and P.Simon. Fabrication of periodic nanostructures by phase-controlled multiple-beam interference. *Applied Physics Letters*, pages 4707–4709, 2003.

68 J.H.Klein-Wiele, J.Bekesi, and P.Simon. Sub-micron patterning of solid materials with ultraviolet femtosecond pulses. *Applied Physics A*, 79:775–778, 2004.

69 B.C. Stuart, M.D. Feit, A.M. Rubenchik, B.W. Shore, and M.D. Perry. Laser-induced damage in dielectrics with nanosecond to subpicosecond pulses. *Phys. Rev. Lett.*, 74:2248–2251, 1995.

70 J. Ihlemann, A. Scholl, H. Schmidt, and B. Wolff-Rottke. Nanosecond and femtosecond excimer-laser ablation of oxide ceramics. *Appl. Phys. A*, 60:411–417, 1995.

71 K. Rubahn and J. Ihlemann. Graded transmission dielectric optical mask by laser ablation. *Appl. Surf. Sci.*, 127 - 129:881–884, 1998.

72 O. Lehmann and M. Stuke. Generation of three-dimensional freestanding metal micro-objects by laser chemical processing. *Appl. Phys. A*, 53:343–345, 1991.

73 D. Bäuerle. *Laser Processing and Chemistry*. Springer-Verlag, Berlin, 1996.

74 K. Rubahn and J. Ihlemann. Private note (1998).

75 S. Mauro, O. Nakamura, and S. Kawata. Three-dimensional microfabrication with two-photon-absorbed photopolymerization. *Opt. Lett.*, 22:132–134, 1997.

76 S. Kawata, H.-B. Sun, T. Tanaka, and K. Takada. Finer features for functional microdevices. *Nature*, 412:697–698, 2001.

77 P. Galajda and P. Ormos. Complex micromachines produced and driven by light. *Appl. Phys. Lett.*, 78:249–251, 2001.

78 Y. Xia and G.M. Whitesides. Soft lithography. *Angew. Chem. Int. Ed.*, 37:550–670, 1998.

79 S.Y. Chou, P.R. Krauss, and P.J. Renstrom. Imprint of sub-25 nm vias and trenches in polymers. *Appl. Phys. Lett.*, 67:3114–3116, 1995.

80 S.Y. Chou, P.R. Krauss, and P.J. Renstrom. Imprint lithography with 25-nanometer resolution. *Science*, 272:85–87, 1996.

81 S.Y. Chou, P.R. Krauss, and P.J. Renstrom. Nanoimprint lithography. *J. Vac. Sci. Techn. B*, 14:4129–4133, 1996.

82 H. Schift, C. David, M. Gabriel, J. Gobrecht, L.J. Heyderman, W. Kaiser, S. Köppel, and L. Scandella. Nanoreplication in polymers using hot embossing and injection molding. *Microelectr. Engin.*, 53:171–174, 2000.

83 H.-C. Scheer, H. Schulz, T. Hoffmann, and C.M. Sotomayor Torres. Problems of the nanoimprinting technique for nanometer scale pattern definition. *J. Vac. Sci. Technol. B*, 16:3917–3921, 1998.

84 G. Engelmann, J. Ziegler, and D. Kolb. Electrochemical fabrication of large arrays of metal nanoclusters. *Surf. Sci.*, 401:L420–L424, 1998.

85 M. Wendel, S. Kühn, H. Lorenz, J.P. Kotthaus, and M. Holland. Nanolithography with an atomic force microscope for integrated fabrication of quantum electronic devices. *Appl. Phys. Lett.*, 65:1775, 1994.

86 M. Wendel, H. Lorenz, and J.P. Kotthaus. Sharpened electron beam deposited tips for high resolution atomic force microscope lithography and imaging. *Appl. Phys. Lett.*, 67:3732, 1995.

87 B.W. Chui, T.D. Stowe, T.W. Kennedy, H.J. Mamin, B.D. Terris, and D. Rugar. Low-stiffness silicon cantilevers for thermal writing and piezoresistive readback with the atomic force microscope. *Appl. Phys. Lett.*, 69:2767–2769, 1996.

88 H.Sugimura and N.Nakagiri. *J.Polym.Sci.Technol.*, 10:661, 1997.

89 W.T. Müller, D.L. Klein, Th. Lee, J. Clarke, P.L. McEuen, and P.G. Schultz. A strategy for the chemical synthesis of nanostructures. *Science*, 268:272–273, 1995.

90 H. Clausen-Schaumann, M. Grandbois, and H.E. Gaub. Enzyme-assisted nanoscale lithography in lipid membranes. *Adv. Mater.*, 10:949–952, 1998.

91 J. Jersch and K. Dickmann. Nanostructure fabrication using laser field enhancement in the near field of a scanning tunneling microscope tip. *Appl. Phys. Lett.*, 68:868–870, 1996.

92 J. Jersch, F. Demming, L. Hildenhagen, and K. Dickmann. Field enhancement of optical radiation in the nearfield of scanning probe microscope tips. *Appl. Phys. A*, 66:29–34, 1998.

93 M.F. Crommie, C.P. Lutz, D.M. Eigler, and E.J. Heller. Waves on a metal surface and quantum corrals. *Surf. Rev. Lett.*, 2:127–137, 1995.

94 D.M. Eigler and E.K. Schweizer. Positioning single atoms with a scanning tunnelling microscope. *Nature*, 344:524–526, 1990.

95 H.C. Manoharan, C.P. Lutz, and D.M. Eigler. Quantum mirages formed by coherent projection of electronic structure. *Nature*, 403:512–515, 2000.

96 J. Kondo. Resistance minimum in dilute magnetic alloys. *Progr. Theor. Phys.*, 32:37–49, 1964.

97 I.V. Markov. *Crystal Growth for Beginners: Fundamentals of Nucleation, Crystal Growth and Epitaxy*. World Scientific, Singapore, 2003.

98 J.W. Matthews, editor. *Epitaxial Growth, Parts A and B*. Academic Press, New York, 1975.

99 F.C. Frank and J.H. van der Merwe. One-dimensional dislocations. i. static theory. *Proc. R. Soc. A*, 198:205–216, 1949.

100 M.Volmer and A.Weber. *Z.Phys.Chem.(Leipzig)*, 119:277, 1926.

101 I.N.Stranski and L.Krastanov. *Sitzungsber.Akad.Wiss.Wien*, 146:797, 1938.

102 J.A. Venables. Atomic processes in crystal growth. *Surf. Sci.*, 299/300:798–817, 1994.

103 K. Reichelt. Nucleation and growth of thin films. *Vacuum*, 38:1083–1099, 1988.

104 H. Brune. Microscopic view of epitaxial metal growth: Nucleation and aggregation. *Surf. Sci. Rep.*, 31:121–229, 1998.

105 P. Jensen, H. Larralde, M. Meunier, and A. Pimpinelli. Growth of three-dimensional structures by atomic deposition on surfaces containing defects: Simulations and theory. *Surface Science*, 412/413:458–476, 1998.

106 H.-G. Rubahn. *Laseranwendungen in der Oberflächenphysik und Materialbearbeitung*. Teubner-Verlag, Stuttgart, 1996.

107 W.M. Tong and R.S. Williams. Kinetics of surface growth: Phenomenology, scaling and mechanisms of smoothening and roughening. *Annu. Rev. Phys. Chem.*, 45:401–438, 1994.

108 M. Rasigni and G. Rasigni. Anomalies of the optical properties of thin lithium layers and their relation to similar anomalies observed with other alkali metals. *J. Opt. Soc. Am.*, 63:775–785, 1973.

109 H. Schmeisser and M. Harsdorff. Investigation of the nucleation of gold on ultra high vacuum cleaved NaCl single crystals. *Z. Naturforsch. A*, 25:1896–1905, 1970.

110 Y. Golan, L. Margulis, and I. Rubinstein. Vacuum deposited gold films. *Surf. Sci.*, 264:312–326, 1992.

111 M. Levlin, A. Laakso, H.E.-M. Niemi, and P. Hautojärvi. Evaporation of gold thin films on mica: effect of evaporation parameters. *Appl. Surf. Sci.*, 115:31–38, 1997.

112 T. Andersson and C.G. Granqvist. Morphology and size distributions of islands in discontinuous films. *J. Appl. Phys.*, 48:1673–1679, 1977.

113 Z.H. Ma, W.D. Sun, I.K. Sou, and G.K.L. Wong. Atomic force microscopy studies of ZnSe self-organized dots fabricated on ZnS/GaP. *Appl. Phys. Lett.*, 73:1340–1342, 1998.

114 J. Bosbach, D. Martin, F. Stietz, T. Wenzel, and F. Träger. Laser-based method for fabricating monodisperse metallic nanoparticles. *Appl. Phys. Lett.*, 74:2605–2607, 1999.

115 L. Ding, J. Li, E. Wang, and S. Dong. K^+ sensors based on supported alkanethiol/phospholipid bilayers. *Thin Solid Films*, 293:153–158, 1997.

116 S. Nowak, T. Pfau, and J. Mlynek. Nanolithography with metastable helium. *Appl. Phys. B*, 63:203–205, 1996.

117 D. Neuschäfer, H. Preiswerk, H. Spahni, E. Konz, and G. Marowsky. Second-harmonic generation using planar waveguides with consideration of pump depletion and absorption. *J. Opt. Soc. Am. B*, 11:649–654, 1994.

118 G. Witte and Ch. Wöll. Growth of aromatic molecules on solid substrates for applications in organic electronics. *J. Mat. Res.*, 19:1889–1916, 2004.

119 Ch. Ziegler. Thin film properties of oligothiophenes. In H.S. Nalwa, editor, *Handbook of Organic Conductive Molecules and Polymers*, volume 3, pages 677–743. John Wiley & Sons Ltd., 1997.

120 B. Krause, A.C. Dürr, K. Ritley, F. Schreiber, H. Dosch, and D. Smilgies. Structure and growth morphology of an archetypal system for organic epitaxy: PTCDA on Ag(111). *Phys. Rev. B*, 66:235404, 2002.

121 M. Brinkmann, S. Graff, C. Straupe, J.-C. Wittmann, C. Chaumont, F. Nuesch, A. Aziz, M. Schaer, and L. Zuppiroli. Orienting tetracene and pentacene thin films onto friction-transferred poly(tetrafluoroethylene) substrate. *J. Phys. Chem. B*, 107:10531–10539, 2003.

122 R. Resel. Crystallographic studies on hexaphenyl thin films - a review. *Thin Solid Films*, 433:1–11, 2003.

123 G.I. Distler. Electrical structure of the surface of crystalline substrates and its influence on nucleation and growth processes. *Kristall und Technik*, 5:73–84, 1970.

124 A.Pockels. *Nature*, 43:437, 1891.

125 K.B. Blodgett and I. Langmuir. Built-up films of barium stearate and their optical properties. *Phys. Rev.*, 51:964–982, 1937.

126 A. Ulman. *An Introduction to Ultrathin Organic Films*. Academic Press, New York, 1991.

127 R. Steiger. Studies of oriented monolayers on solid surfaces by ellipsometry. *Helv. Chim. Acta*, 54:2645–2658, 1971.

128 D. Möbius and H. Bücher. Spectroscopy of monolayer assemblies, part II. In A. Weissberger and B.W. Rossiter, editors, *Physical Methods of Chemistry, Pt. 3b*, New York, 1972. Wiley.

129 S.D. Evans and A. Ulman. Surface potential studies of alkyl-thiol monolayers adsorbed on gold. *Chem. Phys. Lett.*, 170:462–466, 1990.

130 L. Dubois and R.G. Nuzzo. Synthesis, structure, and properties of model organic surfaces. *Annu. Rev. Phys. Chem.*, 43:437–463, 1992.

131 A. Ulman. *Thin Films: Self-Assembled Monolayers of Thiols*. Academic Press, San Diego, 1998.

132 D. Allara and R. Nuzzo. The application of reflection infrared and surface enhanced Raman spectroscopy to the characterization of chemisorbed organic disulfides on Au. *J. Electron. Spectrosc. Relat. Phenom.*, 30:11, 1983.

133 M. Porter, T. Bright, D. Allara, and C. Chidsey. Spontaneous organized molecular assemblies. IV. structural characterization of n-alkyl thiol monolayers on gold by optical ellipsometry, infrared spectroscopy, and electrochemistry. *J. Am. Chem. Soc.*, 109:3559–3568, 1987.

134 N. Tillman, A. Ulman, and T. Penner. Formation of multilayers by self-assembly. *Langmuir*, 5:101–111, 1989.

135 R. Maoz, S. Matlis, E. DiMasi, B.M. Ocko, and J. Sagiv. Self-replicating amphiphilic monolayers. *Nature*, 384:150–153, 1996.

136 A. Ulman and N. Tilman. Self-assembling double layers on gold surfaces: The merging of two chemistries. *Langmuir*, 5:1420–1422, 1989.

137 O. Dannenberger, K. Weiss, H.-J. Himmel, B. Jäger, M. Buck, and Ch. Wöll. An orientation analysis of differently endgroup-functionalised alkanethiols adsorbed on Au substrates. *Thin Solid Films*, 307:183–191, 1997.

138 P. Fenter, A. Eberhardt, and P. Eisenberger. Self-assembly of n-alkyl thiols as disulfides on Au(111). *Science*, 266:1216–1218, 1994.

139 P. Fenter, A. Eberhardt, K.S. Liang, and P. Eisenberger. Epitaxy and chainlength dependent strain in self-assembled monolayers. *J. Chem. Phys.*, 106:1600–1608, 1997.

140 P. Fenter, F. Schreiber, L. Berman, G. Scoles, P. Eisenberger, and M.J. Bedzyk. On the structure and evolution of the buried S/Au interface in self-assembled monolayers: X-ray standing wave results. *Surface Science*, 412/413:213–235, 1998.

141 L. Strong and G.M. Whitesides. Structures of self-assembled monolayer films of organosulfur compounds adsorbed on gold single crystals: Electron diffraction studies. *Langmuir*, 4:546–558, 1988.

142 B. Heinz and H. Morgner. MIES investigation of alkanethiol monolayers self-assembled on Au(111) and Ag(111) surfaces. *Surface Science*, 372:100–116, 1997.

143 R. Gerlach, G. Polanski, and H.-G. Rubahn. Structural manipulation of ultrathin organic films on metal surfaces: the case of decane thiol/Au(111). *Appl. Phys. A*, 65:375–377, 1997.

144 N. Camillone, T. Leung, and G. Scoles. A low energy helium atom diffraction study of decanethiol self-assembled on Au(111). *Surf. Sci.*, 373:333–349, 1997.

145 M. Flörsheimer, A.J. Steinfort, and P. Günter. Lattice constants of Langmuir–Blodgett films measured by atomic force microscopy. *Surf. Sci. Lett.*, 297:L39–L42, 1993.

146 E. Delamarche and B. Michel. Structure and stability of self-assembled monolayers. *Thin Solid Films*, 273:54–60, 1996.

147 M. Toerker, R. Staub, T. Fritz, T. Schmitz-Hübsch, F. Sellam, and K. Leo. Annealed decanethiol monolayers on Au(111) - intermediate phases between structures with high and low molecular surface density. *Surf. Sci.*, 445:100–108, 2000.

148 E. Delamarche, B. Michel, H. A. Biebuyck, and C. Gerber. Golden interfaces: The surface of self-assembled monolayers. *Adv. Mater.*, 8:719–729, 1996.

149 M. Buck, F. Eisert, M. Grunze, and F. Träger. Second-order nonlinear susceptibilities of surfaces: A systematic study of the wavelength and coverage dependence of thiol adsorption on polycrystalline gold. *Appl. Phys. A*, 60:1–12, 1995.

150 F. Schreiber. Structure and growth of self-assembling monolayers. *Progr. Surf. Sci.*, 65:151–257, 2000.

151 R. Staub, M. Toerker, T. Fritz, T. Schmitz-Hübsch, F. Sellam, and K. Leo. Flat lying pin-stripe phase of decanethiol self-assembled monolayers on Au(111). *Langmuir*, 14:6693–6698, 1998.

152 D.R. Jung and A.W. Czanderna. Chemical and physical interactions at metal/self-assembled organic monolayer interfaces. *Crit. Rev. Sol. State Mat. Sci.*, 19:1–54, 1994.

153 R.O.Al-Kaysi and C.J.Bardeen. *Chem.Commun.*, pages 1224–1226, 2006.

154 J.-H.Fuhrhop, U.Bindig, and U.Siggel. *J.Am.Chem.Soc.*, 115:11036–11037, 1993.

155 A.D.Schwab, D.E.Smith, C.S.Rich, E.R.Young, W.F.Smith, and J.C.DePaula. *J.Phys.Chem.B*, 107:11339–11345, 2003.

156 R.Rotomskis, R.Augulis, V.Snitka, R.Valiokas, and B.Liedberg. *J.Phys.Chem.B*, 108:2833–2838, 2004.

157 K.Takazawa, Y.Kitahama, Y.Kimuar, and G.Kido. *Nano Lett.*, 5:1293–1296, 2005.

158 A.N.Lebedenko, G.Y.Guralchuk, A.V.Sorokin, S.L.Yefimova, and Y.V.Malyukin. *J.Phys.Chem.B*, 110:17772–17775, 2006.

159 A.P.H.J.Schenning and E.W.Meijer. Supramolecular electronics: Nanowires from self-assembled pi-conjugated systems. *Chem.Commun.*, pages 3245–3258, 2005.

160 R.F.Service. *Science*, 312:1593–1594, 2006.

161 D.Appell. *Nature*, 419:554–555, 2002.

162 Y.Xia, P.Yag, Y.Sun, Y.Wu, B.Mayers, B.Gates, Y.Yin, F.Kim, and H.Yan. One-dimensional nanostructures:synthesis, characterization and applications. *Adv.Mater.*, 15:353–389, 2003.

163 M.Law, J.Goldberger, and P.Yang. *Annju.Rev.Mat.Res.*, 34:83–122, 2004.

164 D.J.Sirbuly, M.Law, H.Yan, and P.Yang. *J.Phys.Chem.B*, 109:15190–15213, 2005.

165 P.J.Pauzauskie and P.Yang. *Mat.Today*, 9:36–45, 2006.

166 A.Javey, S.Nam, R.S.Friedmann, H.Yan, and C.M.Lieber. Layer-by-layer assembly of nanowires for three-dimensional, multifunctional electronics. *Nano Lett.*, 7:773–777, 2007.

167 F. Balzer and H.-G. Rubahn. Dipole-assisted self-assembly of light-emitting p-*n*P needles on mica. *Appl. Phys. Lett.*, 79:3860–3862, 2001.

168 F. Balzer and H.-G. Rubahn. Growth control and optics of organic nanoaggregates. *Adv. Funct. Mater.*, 15:17–24, 2005.

169 P. Puschnig and C. Ambrosch-Draxl. Density-functional study of the oligomers of poly(*para*-phenylene): Band structures and dielectric tensors. *Phys. Rev. B*, 60:7891–7898, 1999.

170 F. Balzer, V.G. Bordo, A.C. Simonsen, and H.-G. Rubahn. Optical waveguiding in individual nanometer-scale organic fibers. *Phys. Rev. B*, 67:115408, 2003.

171 F. Balzer and H.-G. Rubahn. Laser-controlled growth of needle-shaped organic nanoaggregates. *Nano Letters*, 2:747–750, 2002.

172 J. Brewer, C. Maibohm, L. Jozefowski, L. Bagatolli, and H.-G. Rubahn. A 3D view on free-floating, space-fixed and surface-bound *para*-phenylene nanofibres. *Nanotechnology*, 16:2396–2401, 2005.

173 M. Schiek, A. Lützen, R. Koch, K. Al-Shamery, F. Balzer, R. Frese, and H.-G. Rubahn. Nanofibers from functionalized *para*-phenylene molecules. *Appl. Phys. Lett.*, 86:153107, 2005.

174 J.Kjelstrup-Hansen, H.H.Henrichsen, P.Bogild, and H.-G.Rubahn. *Thin Solid Films*, 515:827, 2006.

175 J.Brewer, M.Schiek, A.Luetzen, K.Al-Shamery, and H.-G.Rubahn. *Nano Lett.*, 6:2656, 2006.

176 K.Al-Shamery, H.-G.Rubahn, and H.Sitter, editors. *Organic Nanostructures for Next Generation Devices*, volume 101 of *Springer Series in Materials Science*. Springer, Berlin, 2008.

177 M. Zinke-Allmang, L.C. Feldman, and M.H. Grabow. Clustering on surfaces. *Surf. Sci. Rep.*, 16:377–463, 1992.

178 R. Pascal, C. Zarnitz, M. Bode, and R. Wiesendanger. Fabrication of atomic gratings based on self-organization of adsorbates with repulsive interaction. *Appl. Phys. A*, 65:81–83, 1997.

179 H. Roeder, K. Bromann, H. Brune, and K. Kern. Strain mediated two-dimensional growth kinetics in metal heteroepitaxy. *Surface Science*, 376:13–31, 1997.

180 H. Brune, M. Giovannini, K. Bromann, and K. Kern. Self-organized growth of nanostructure arrays on strain-relief patterns. *Nature*, 394:451–452, 1998.

181 K. Pohl, M.C. Bartelt, J. de la Figuera, N.C. Bartelt, J.Hrbek, and R.Q. Hwang. Identifying the forces responsible for self-organization of nanostructures at crystal surfaces. *Nature*, 397:238–241, 1999.

182 H.-J. Ernst, F. Fabre, R. Folkerts, and J. Lapujoulade. Observation of a growth instability during low temperature molecular beam epitaxy. *Phys. Rev. Lett.*, 72:112–115, 1994.

183 J.-K. Zuo and J.-F. Wendelken. Evolution of mound morphology in reversible homoepitaxy on Cu(100). *Phys. Rev. Lett.*, 78:2791–2794, 1997.

184 J.A. Stroscio, D.T. Pierce, M.D. Stiles, A. Zangwill, and L.M. Sander. Coarsening of unstable surface features during Fe(001) homoepitaxy. *Phys. Rev. Lett.*, 75:4246–4249, 1995.

185 G. Ehrlich and F.G. Hudda. Atomic view of surface self-diffusion: Tungsten on tungsten. *J. Chem. Phys.*, 44:1039, 1966.

186 R.L. Schwoebel. Step motion on crystal surfaces. II. *J. Appl. Phys.*, 40:614, 1969.

187 R. Gerlach, T. Maroutian, L. Douillard, D. Martinotti, and H.-J. Ernst. A novel method to determine the Ehrlich-Schwoebel barrier. *Surf. Sci.*, 480:97–102, 2001.

188 C. Zeng, B. Wang, B. Li, H. Wang, and J.G. Hou. Self-assembly of one-dimensional molecular and atomic chains using striped alkanethiol structures as templates. *Appl. Phys. Lett.*, 79:1685, 2001.

189 S. Lukas, G. Witte, and Ch. Wöll. Novel mechanism for molecular self-assembly on metal substrates: Unidirectional rows of pentacene on Cu(110) produced by a substrate-mediated repulsion. *Phys. Rev. Lett.*, 88:028301–1–028301–4, 2002.

190 A.B. Stiles, editor. *Catalyst Supports and Supported Catalysts: Theoretical and Applied Concepts*. Butterworth, Boston, 1987.

191 M. Frank, S. Andersson, J. Libuda, S. Stempel, A. Sandell, B. Brena, A. Giertz, P.A. Brühwiler, M. Bäumer, N. Mårtensson, and H.-J. Freund. Particle size dependent CO dissociation on alumina-supported Rh: A model study. *Chem. Phys. Lett.*, 279:92–99, 1997.

192 K. Bromann, Ch. Félix, H. Brune, W. Harbich, R. Monot, J. Buttet, and K. Kern. Controlled deposition of size-selected silver nanoclusters. *Science*, 274:956–958, 1996.

193 P. Jensen. Growth of nanostructures by cluster deposition: Experiments and simple models. *Rev. Mod. Phys.*, 71:1695–1735, 1999.

194 C. Binns. Nanoclusters deposited on surfaces. *Surf. Sci. Rep.*, 44:1–49, 2001.

195 R. Neuendorf. Private note (2003).

196 R. Neuendorf, R.E. Palmer, and R. Smith. Low energy deposition of size-selected Si clusters onto graphite. *Chem. Phys. Lett.*, 333:304–307, 2001.

197 A. Hilger, M. Tenfelde, and U. Kreibig. Silver nanoparticles deposited on dielectric surfaces. *Appl. Phys. B*, 73:361–372, 2001.

198 J.H. Weaver and G.D. Waddill. Cluster assembly of interfaces: Nanoscale engineering. *Science*, 251:1444–1451, 1991.

199 L. Huang, S.J. Chey, and J.H. Weaver. Buffer-layer-assisted growth of nanocrystals: Ag-Xe-Si(111). *Phys. Rev. Lett.*, 80:4095–4098, 1998.

200 S.J. Chey, L. Huang, and J.H. Weaver. Self-assembly of multilayer arrays from Ag nanoclusters delivered to

Ag(111) by soft landing. *Surf. Sci.*, 419:L100–L106, 1998.

201 G. Schmid, R. Pfeil, R. Boese, F. Bandermann, S. Meyer, G.H.M. Calis, and J.W.A. van der Velden. $Au_{55}[P(C_6H_5)_3]_{12}CI_6$ - ein Goldcluster ungewöhnlicher Größe. *Chem. Ber.*, 114:3634–3642, 1981.

202 D. Babonneau, T. Cabioc'h, A. Naudon, J.C. Girard, and M.F. Denanot. Silver nanoparticles encapsulated in carbon cages obtained by co-sputtering of the metal and graphite. *Surf. Sci.*, 409:358–371, 1998.

203 A.J. Parker, P.A. Childs, and R.E. Palmer. Deposition of passivated gold nanoclusters onto prepatterned substrates. *Appl. Phys. Lett.*, 74:2833, 1999.

204 N.Hallas. The optical properties of nanoshells. *Optics and Photonics News*, 8:26–31, 2002.

205 J. Walls and R. Smith. *Surface Science Techniques*. Pergamon Press, Oxford, 1994.

206 J. Yates, Jr. *Experimental Innovations in Surface Science: A Guide to Practical Laboratory Methods and Instruments*. Springer-Verlag, New York, 1998.

207 E. Abbe. Beiträge zur Theorie des Mikroskops und der mikroskopischen Wahrnehmung. *Arch. Mikrosk. Anat.*, 9:413–468, 1873.

208 Lord Rayleigh. Investigations in optics with special reference to the spectroscope. *Philos. Mag.*, 8:261– 274, 403–411, 477–486, 1879.

209 C.S.Nachet. *Compte Rendu de l'Academie des Sciences*, XXIV:976, 1847.

210 P.A. Temple. Total internal reflection microscopy: A surface inspection technique. *Appl. Opt.*, 20:2656, 1981.

211 F. Balzer, V.G. Bordo, R. Neuendorf, K. Al-Shamery, A.C. Simonsen, and H.-G. Rubahn. Organic nanoaggregates: A window to submicron optics. *IEEE Trans. Nanotechn.*, 3:67–72, 2004.

212 F.Zernicke. Nobel price lecture. *The Nobel Foundation*, 1953.

213 Q. Wu, R.D. Grober, D. Gammon, and D.S. Katzer. Excitons, biexcitons, and electron-hole plasma in a narrow 2.8-nm GaAs/Al_xGa_{1-x}As quantum well. *Phys. Rev. B*, 62:13022–13027, 2000.

214 T. Wilson. *Confocal Microscopy*. Academic, London, 1990.

215 M. Schrader, S.W. Hell, and H.T.M. van der Voort. Three-dimensional super-resolution with a 4pi-confocal microscope using image restoration. *J. Appl. Phys.*, 84:4033–4042, 1998.

216 C.J.R. Sheppard. Image formation in three-photon fluorescence microscopy. *Bioimaging*, 4:124–128, 1996.

217 L.A. Bagatolli, S. Sanchez, T. Hazlett, and E. Gratton. Giant vesicles, laurdan, and two-photon fluorescence microscopy: Evidence of lipid lateral separation in bilayers. In G. Marriot and I. Parker, editors, *Methods in Enzymology - Biophotonics Part A*, pages 481–500, San Diego, 2003. Academic Press.

218 L.A. Bagatolli. University of Southern Denmark, private communication.

219 T.A. Klar and S.W. Hell. Subdiffraction resolution in far-field fluorescence microscopy. *Opt. Lett.*, 24:954–956, 1999.

220 K.I.Willig, S.O.Rizzoli, V.Westphal, R.Jahn, and S.W.Hell. STED microscopy reveals that synaptotagmin remains clustered after synaptic vesicle exocytosis. *Nature*, 440:935–939, 2006.

221 D. Hoenig and D. Moebius. Direct visualization of monolayers at the air-water interface by Brewster angle microscopy. *J. Phys. Chem.*, 95:4590–4592, 1991.

222 S. Henon and J. Meunier. Microscope at the Brewster angle: Direct observation of first-order phase transitions in monolayers. *Rev. Sci. Instr.*, 62:936–939, 1991.

223 M.N.G. de Mul and J. Adin Mann, Jr. Determination of the thickness and optical properties of a Langmuir film from the domain morphology by Brewster angle microscopy. *Langmuir*, 14:2455–2466, 1998.

224 H. Alexander. *Physikalische Grundlagen der Elektronenmikroskopie*. Teubner, Stuttgart, 2002.

225 O.Scherzer. *Z.Phys.*, 101:593, 1936.

226 J.M. Cowley. Twenty forms of electron holography. *Ultramicroscopy*, 41:335–348, 1992.

227 A. Orchowski, W.D. Rau, and H. Lichte. Electron holography surmounts resolution limit of electron microscopy. *Phys. Rev. Lett.*, 74:399–402, 1995.

228 E. Müller. Das Feldionenmikroskop. *Z. Phys.*, 131:136–142, 1951.

229 C.J. Chen. *Introduction to Scanning Tunneling Microscopy*. Oxford University Press, Oxford, 1993.

230 H. Fuchs. SXM-Methoden – nützliche Werkzeuge für die Praxis? *Phys. Blätter*, 50:837–843, 1994.

231 G. Binnig, H. Rohrer, C. Gerber, and E. Weibel. Surface studies by scanning tunneling microscopy. *Phys. Rev. Lett.*, 49:57–61, 1982.

232 T.R. Hicks and P.D. Atherton. *The NanoPositioning Book*. Queensgate Instruments Limited, Berkshire, 1997.

233 Firma Nanosensors. Wetzlar, 1999.

234 C. Bai. *Scanning Tunneling Microscopy and Its Application*, volume 32 of *Springer Series in Surface Science*. Springer, Berlin, 2000.

235 M.S. Hoogeman, D. Glastra van Loon, R.W.M. Loos, H.G. Ficke, E. de Haas, J.J. van der Linden, H. Zeijlemaker, L. Kuipers, M.F. Chang, M.A.J. Klik, and J.W.M. Frenken. Design and performance of a programmable-temperature scanning tunneling microscope. *Rev. Sci. Instrum.*, 69:2072, 1998.

236 L. Petersen, M. Schunack, B. Schaefer, T.R. Linderoth, P.B. Rasmussen, P.T. Sprunger, E. Laegsgaard, I. Stensgaard, and F. Besenbacher. A fast-scanning, low- and variable-temperature scanning tunneling microscope. *Rev. Sci. Instrum.*, 72:1438, 2001.

237 E. Laegsgaard, L. Österlund, P. Thostrup, P.B. Rasmussen, I. Stensgaard, and F. Besenbacher. A high-pressure scanning tunneling microscope. *Rev. Sci. Instrum.*, 72:3537, 2001.

238 F. Besenbacher. Scanning tunnelling microscopy studies of metal surfaces. *Rep. Prog. Phys.*, 59:1737–1802, 1996.

239 G. Binnig, C.F. Quate, and C. Gerber. Atomic force microscope. *Phys. Rev. Lett.*, 56:930–933, 1986.

240 S. Magonov and M.-H. Whangbo. *Surface Analysis with STM and AFM*. VCH, Weinheim, 1996.

241 B.A. Todd and S.J. Eppell. A method to improve the quantitative analysis of SFM images at the nanoscale. *Surf. Sci.*, 491:473–483, 2001.

242 V.J. Morris, A.P. Gunning, and A.R. Kirby. *Atomic Force Microscopy for Biologists*. Imperial College Press, London, 1999.

243 R. Wiesendanger. *Scanning Probe Microscopy and Spectroscopy: Methods and Applications*. Cambridge University Press, Cambridge, 1994.

244 D.Sarid. *Scanning Force Microscopy: With Applications to Electric, Magnetic and Atomic Forces*, volume 5 of *Oxford Series in Optical and Imaging Sciences*. Oxford University Press, Oxford, 1994.

245 R. Wiesendanger, editor. *Scanning Probe Microscopy: Analytical Methods*, Berlin, 1998. Springer.

246 E.H. Synge. Method for extending microscopic radiation into the ultramicroscopic region. *Philos. Mag.*, 6:356–358, 1928.

247 E.A. Ash and G. Nicholls. Superresolution aperture scanning microscope. *Nature*, 237:510–512, 1972.

248 D.W. Pohl. Optical near-field microscope, 1982. European Patent Application No. 0112401, 27. Dez. 1982.

249 D.W. Pohl, W. Denk, and M. Lanz. Optical stethoscopy: Image recording with resolution $\lambda/20$. *Appl. Phys. Lett.*, 44:651–653, 1984.

250 D.W. Pohl and D. Courjon. *Near Field Optics*. Kluwer, Dordrecht, 1993.

251 R.C. Reddick, R.J. Warmack, and T.L. Ferrell. New form of scanning optical microscopy. *Phys. Rev. B*, 39:767–770, 1989.

252 M.A. Paesler and P.J. Moyer. *Near-Field Optics*. John Wiley, New York, 1996.

253 B. Hecht, B. Sick, U.P. Wild, V. Deckert, R. Zenobi, O.J.F. Martin, and D.W. Pohl. Scanning near-field optical microscopy with aperture probes: Fundamentals and applications. *J. Chem. Phys.*, 112:7761–7774, 2000.

254 H. Sturm, F. Balzer, and H.-G. Rubahn. Bundesanstalt für Materialwissenschaften, Berlin, 2003. unpublished.

255 V. Volkov, S. I. Bozhevolnyi, V.G. Bordo, and H.-G. Rubahn. Near field imaging of organic nanofibres. *J. Microscopy*, 215:241–244, 2004.

256 F. Zenhausern, M.P. O'Boyle, and H.K. Wickramasinghe. Apertureless near-field optical microscope. *Appl. Phys. Lett.*, 65:1623–1625, 1994.

257 U. Fischer and D. Pohl. Observation of single-particle plasmons by near-field optical microscopy. *Phys. Rev. Lett.*, 62:458–461, 1989.

258 J. Erland, S.I. Bozhevolnyi, K. Pedersen, J.R. Jensen, and J.M. Hvam. Second-harmonic imaging of semiconductor quantum dots. *Appl. Phys. Lett.*, 77:806–808, 2000.

259 B. Vohnsen, S.I. Bozhevolnyi, K. Pedersen, J. Erland, J.R. Jensen, and J.M. Hvam. Second-harmonic scanning optical microscopy of semiconductor quantum dots. *Opt. Commun.*, 189:305–311, 2001.

260 J. Beermann, S.I. Bozhevolnyi, V.G. Bordo, and H.-G. Rubahn. Two-photon mapping of local molecular orientations in hexaphenyl nanofibers. *Opt. Comm.*, 237:423–429, 2004.

261 J. Michaelis, C. Hettich, J. Mlynek, and V. Sandoghdar. Optical microscopy using a single-molecule light source. *Nature*, 405:325–328, 2000.

262 V.S. Letokhov. Possible laser modification of field-ion microscopy. *Phys. Lett. A*, 51:231–232, 1975.

263 S.V. Chekalin, V.S. Letokhov, V.S. Likhachev, and V.G. Movshev. Laser photoion projector. *Appl. Phys. B*, 33:57–61, 1984.

264 G. T. Shubeita, S.K. Sekatskii, M. Chergui, G. Dietler, and V.S. Letokhov. Investigation of nanolocal fluorescence resonance energy transfer for scanning probe microscopy. *Appl. Phys. Lett.*, 74:3453–3455, 1999.

265 W. Krieger, A. Hornsteiner, E. Soergel, C. Sammet, M. Volcker, and H. Walther. Laser-driven scanning tunneling microscope. *Laser Phys.*, 6:334–338, 1996.

266 C.J. Hood, T.W. Lynn, A.C. Doherty, A.S. Parkins, and H.J. Kimble. The atom-cavity microscope: Single atoms bound in orbit by single photons. *Science*, 287:1447–1453, 2000.

267 G.J.Burch. *Nature*, 54:111, 1896.

268 H. Wolter. Spiegelsysteme streifenden Einfalls als abbildende Optiken für Röntgenstrahlen. *Ann. Phys.*, 445:94–114, 1952.

269 A.V. Baez. Study in diffraction microscopy with special reference to x-rays. *J. Opt. Soc. Am.*, 42:756, 1952.

270 G.Schmahl and D.Rudolph. *Optik*, 29:577, 1969.

271 G. Schmahl. X-ray optics. In H. Niedrig, editor, *Optics of Waves and Particles*, page 933, Berlin, 1999. Walter de Gruyter.

272 R.B. Doak. In E. Hulpke, editor, *He-Scattering: A Gentle and Sensitive Tool in Surface Science*, Berlin, 1992. Springer.

273 B. Holst and W. Allison. An atom-focusing mirror. *Nature*, 390:244–244, 1997.

274 A. Watson. Helium beam shows the gentle, sensitive touch. *Science*, 286:1831, 1999.

275 R.J. Wilson, B. Holst, and W. Allison. Optical properties of mirrors for focusing of non-normal incidence atom beams. *Rev. Sci. Instrum.*, 70:2960–2967, 1999.

276 D.A. MacLaren, W. Allison, and B. Holst. Single crystal optic elements for helium atom microscopy. *Rev. Sci. Instrum.*, 71:2625–2634, 2000.

277 S. Rehbein, R.B. Doak, R.E. Grisenti, G. Schmahl, J.P. Toennies, and Ch. Wöll. Nanostructuring of zone plates for helium atom beam focusing. *Microelectr. Eng.*, 53:685–688, 2000.

278 O. Carnal, M. Sigel, T. Sleator, H. Takuma, and J. Mlynek. Imaging and focusing of atoms by a fresnel zone plate. *Phys. Rev. Lett.*, 67:3231–3234, 1991.

279 R.B. Doak, R.E. Grisenti, S. Rehbein, G. Schmahl, J.P. Toennies, and Ch. Wöll. Towards realization of an atomic de Broglie microscope: Helium atom focusing using Fresnel zone plates. *Phys. Rev. Lett.*, 83:4229–4232, 1999.

280 R.E. Grisenti, W. Schöllkopf, J.P. Toennies, G.C. Hegerfeldt, and T. Köhler. Determination of atom-surface van der Waals potentials from transmission-grating diffraction intensities. *Phys. Rev. Lett.*, 83:1755–1758, 1999.

281 R.E. Grisenti, W. Schöllkopf, J.P. Toennies, G.C. Hegerfeldt, T. Köhler, and M. Stoll. Determination of the bond length and binding energy of the helium dimer by diffraction from a transmission grating. *Phys. Rev. Lett.*, 85:2284–2287, 2000.

282 R.E. Grisenti, W. Schöllkopf, J.P. Toennies, J.R. Manson, T.A. Savas, and H.I. Smith. He-atom diffraction from nanostructure transmission gratings: The role of imperfections. *Phys. Rev. A*, 61:033608, 2000.

283 I. Bloch, M. Köhl, M. Greiner, T.W. Hänsch, and T. Esslinger. Optics with an atom laser beam. *Phys. Rev. Lett.*, 87:030401, 2001.

284 W.Demtroeder. *Laser Spectroscopy: Basic Principles*, volume 1. Springer, Berlin, 4th edition, 2007.

285 S. Brasselet and W.E. Moerner. Fluorescence behavior of single-molecule pH-sensors. *Single Mol.*, 1:17–23, 2000.

286 J.E. Sipe. New Green-function formalism for surface optics. *J. Opt. Soc. Am. B*, 4:481–489, 1987.

287 F. Balzer. *Lineare und Nichtlineare Optik an Alkali-Inselfilmen*. Ph.D. Thesis, University of Göttingen, 1998.

288 V. Mizrahi and J.E. Sipe. Phenomenological treatment of surface second-harmonic generation. *J. Opt. Soc. Am. B*, 5:660–667, 1988.

289 J.D. Jackson. *Classical Electrodynamics*. John Wiley & Sons, New York, 1999.

290 F. Balzer, S.D. Jett, and H.-G. Rubahn. Alkali cluster films on insulating substrates: Comparison between scanning force microscopy and extinction data. *Chem. Phys. Lett.*, 297:273–280, 1998.

291 M. Raether. *Surface Plasmons on Smooth and Rough Surfaces and on Gratings*, volume 111 of *Springer Tracts in Modern Physics*. Springer-Verlag, Berlin, 1988.

292 H. Ditlbacher, J.R. Krenn, N. Felidj, B. Lamprecht, G. Schider, M. Salerno, A. Leitner, and F.R. Aussenegg. Fluorescence imaging of surface plasmon fields. *Appl. Phys. Lett.*, 80:404–406, 2002.

293 A. Bouhelier, T. Huser, H. Tamaru, H.-J. Güntherodt, D.W. Pohl, F.I. Baida, and D. Van Labeke. Plasmon optics of structured silver films. *Phys. Rev. B*, 63:155404, 2001.

294 E. Kretschmann and H. Raether. Radiative decay of non-radiative surface plasmons excited by light. *Z. Naturforsch.*, 23A:2135–2136, 1968.

295 L. Jozefowski, V.V. Petrunin, and H.-G. Rubahn. unpublished (2001).

296 J. Tominaga, C. Mihalcea, D. Büchel, H. Fukuda, T. Nakano, N. Atoda, H. Fuji, and T. Kikukawa. Local plasmon photonic transistor. *Appl. Phys. Lett.*, 78:2417, 2001.

297 H.G. Tompkins. *A User's Guide to Ellipsometry*. Dover, 2006.

298 Ch. Jung, O. Dannenberger, Y. Xu, M. Buck, and M. Grunze. Self-assembled monolayers from organosulfur compounds: A comparison between sulfides, disulfides, and thiols. *Langmuir*, 14:1103–1107, 1998.

299 A.P.F. Turner, I. Karube, and G.S. Wilson. *Biosensors: Fundamentals and Applications*. Oxford University Press, Oxford, 1987.

300 M. Chevrollier, M. Fichet, M. Oria, G. Rahmat, D. Bloch, and M. Ducloy. High resolution selective reflection spectroscopy as a probe of long-range surface interaction: Measurement of the surface van der Waals attraction on excited Cs atoms. *J. Phys. II France*, 2:631–657, 1992.

301 V.G. Bordo and H.-G. Rubahn. Two-photon evanescent-wave spectroscopy of alkali-metal atoms. *Phys. Rev. A*, 60:1538–1548, 1999.

302 V.G. Bordo, J. Loerke, and H.-G. Rubahn. Two-photon evanescent-volume wave spectroscopy: A new account of gas-solid dynamics in the boundary layer. *Phys. Rev. Lett.*, 86:1490–1493, 2001.

303 H.-G. Rubahn. *Laser Applications in Surface Science and Technology*. Wiley, Chichester, 1999.

304 F. de Fornel. *Evanescent Waves. From Newtonian Optics to Atomic Optics*. Springer, Berlin, 2001.

305 D.Courjon. *Near-Field Microscopy and Near-Field Optics*. Imperial College Press, London, 2003.

306 B. Koopmans, F. van der Woude, and G.A. Sawatzky. Surface symmetry resolution of nonlinear optical techniques. *Phys. Rev. B*, 46:12780–12783, 1992.

307 A. Yariv. *Quantum electronics*. John Wiley & Sons, New York, 3. edition, 1989.

308 J. McGilp. Determining metal-semiconductor interface structure by optical second-harmonic-generation. *J. Vac. Sci. Technol. A*, 5:1442–1446, 1987.

309 R. Bavli, D. Yogev, S. Efrima, and G. Berkovic. Second harmonic generation studies of silver metal liquidlike films. *J. Phys. Chem.*, 95:7422–7426, 1991.

310 G. Berkovic and S. Efrima. Second harmonic generation from composite films of spheroidal metal particles. *Langmuir*, 9:355–357, 1993.

311 F. Balzer and H.-G. Rubahn. Interference effects in the optical second harmonic generation from ultrathin alkali films. *Opt. Comm.*, 185:493–499, 2000.

312 R.A. Fisher, editor. *Optical phase conjugation*. Academic Press, New York, 1983.

313 E. Stelzer, H. Ruf, and E. Grell. In E.O. Schulz-DuBois, editor, *Photon Correlation Techniques in Fluid Mechanics*, Berlin, 1982. Springer.

314 M. van Hove, W. Weinberg, and C.-M. Chan. *Low-Energy Electron Diffraction*. Springer-Verlag, Berlin, 1986.

315 E. Bauer. Low energy electron reflection microscopy. In S.S. Breese, Jr., editor, *Fifth International Congress for Electron Microscopy*, volume 1, New York, 1962. Academic Press.

316 E. Bauer. LEEM basics. *Surf. Rev. Lett.*, 5:1275–1286, 1998.

317 H. Ibach and D.L. Mills. *Electron Energy Loss Spectroscopy and Surface Vibrations*. Academic Press, New York, 1982.

318 W. Kress and F.D. Wette, editors. *Surface Phonons*, Berlin, 1991. Springer-Verlag.

319 E. Hulpke, editor. *Helium Atom Scattering from Surfaces*, Berlin, 1992. Springer-Verlag.

320 J.P. Toennies and K. Winkelmann. Theoretical studies of highly expanded free jets: Influence of quantum effects and a realistic intermolecular potential. *J. Chem. Phys.*, 66:3965–3979, 1977.

321 G. Benedek and J.P. Toennies. Helium atom scattering spectroscopy of surface phonons: Genesis and achievements. *Surf. Sci.*, 299 - 300:587–611, 1994.

322 J. Jensen, P. Morgen, and S. Tougaard. 2002. Private note (2002).

323 H.S. Nalwa, editor. *Handbook of Organic-Inorganic Hybrid Materials and Nanocomposites*, volume 1 & 2. American Scientific Publishers, Stevenson Ranch, CA, 2003.

324 M.T. Cuberes, J.K. Gimzewski, and R.R.Schlittler. Room-temperature repositioning of individual C_{60} molecules at Cu steps: Operation of a molecular counting device. *Appl. Phys. Lett.*, 69:3016–3018, 1996.

325 R.M. Nyffenegger and R.M. Penner. Nanometer-scale surface modification using the scanning probe microscope: Progress since 1991. *Chem. Rev.*, 97:1195–1230, 1997.

326 M. Giersig and P. Mulvaney. Preparation of ordered colloid monolayers by electrophoretic deposition. *Langmuir*, 9:3408–3413, 1993.

327 H. Haberland, editor. *Clusters of Atoms and Molecules I and II*. Springer, Berlin, 1994.

328 G. Schmid. *Clusters and Colloids. From Theory to Applications*. Wiley-VCH, Weinheim, 1998.

329 E. Lugovoj, J.P. Toennies, and A. Vilesov. Manipulating and enhancing chemical reactions in Helium droplets. *J. Chem. Phys.*, 112:8217, 2000.

330 J.P. Toennies, A.F. Vilesov, and K.B. Whaley. Superfluid helium droplets: An ultracold nanolaboratory. *Phys. Today*, 54(2):31–37, 2001.

331 J.P. Toennies and A.F. Vilesov. Spectroscopy of atoms and molecules in liquid helium. *Annu. Rev. Phys. Chem.*, 49:1–41, 1998.

332 R.P. Andres, S. Datta, D.B. Janes, C.P. Kubiak, and R. Reifenberger. The design, fabrication, and electronic properties of self-assembled molecular nanostructures. In H. Nalwa, editor, *The Handbook of Nanostructured Materials and Nanotechnology*, pages 180–232, San Diego, 1998. Academic Press.

333 K.V. Sarathy, P.J. Thomas, G.U. Kulkarni, and C.N.R. Rao. Superlattices of metal and metal-semiconductor quantum dots obtained by layer-by-layer deposition of nanoparticle arrays. *J. Phys. Chem. B*, 103:399–401, 1999.

334 F.Voegtle. *Supramolekulare Chemie*. Teubner, Stuttgart, 1992.

335 P.F.H. Schwab, M.D. Levin, and J. Michl. Molecular rods. 1. simple axial rods. *Chem Rev.*, 99:1863–1934, 1999.

336 A.J. Berresheim, M. Müller, and K. Müllen. Polyphenylene nanostructures. *Chem. Rev.*, 99:1747–1786, 1999.

337 J.M. Tour. *Chem.Rev.*, 96:537, 1996.

338 V.R. Thalladi, R. Boese, S. Brasselet, I. Ledoux, J. Zyss, R.K.R. Jetti, and G.R. Desiraju. Steering noncentrosymmetry into the third dimension: Crystal engineering of an octupolar nonlinear optical crystal. *Chem. Commun.*, pages 1639–1640, 1999.

339 G.R. Newkome, E. He, and C.N. Moorefield. Suprasupermolecules with novel properties: Metallodendrimers. *Chem. Rev.*, 99:1689–1746, 1999.

340 H. Le Bozec, T. Le Bouder, O. Maury, A. Bondon, I. Ledoux, S. Deveau, and J. Zyss. Supramolecular octupolar self-ordering towards nonlinear optics. *Adv. Mater.*, 13:1677–1681, 2001.

341 R.R. Schlittler, J.W. Seo, J.K. Gimzewski, C. Durkan, M.S.M. Saifullah, and M.E. Welland. Single crystals of single-walled carbon nanotubes formed by self-assembly. *Science*, 292:1136–1139, 2001.

342 Y.J. Chabal, editor. *Fundamental Aspects of Silicon Oxidation*. Springer Series in Materials Science. Springer, Berlin, 2001.

343 J. Dabrowski and H.-J. Müssig, editors. *Silicon Surfaces and Formation of Interfaces: Basic Science in the Industrial World*. World Scientific, River Edge, 2000.

344 P. Morgen, F.K. Dam, C. Gundlach, T. Jensen, L.-B. Taekker, S. Tougaard, and K. Pedersen. In *Recent Research Developments in Applied Physics*. TransWorld Research Publishers, 2002.

345 P. Morgen, T. Jensen, C. Gundlach, L.-B. Tække r, S.V. Hoffman, and K. Pedersen. From oxygen adsorption to the growth of thin oxides on silicon surfaces. *Comp. Mat. Sci.*, 21:481–487, 2001.

346 P. Morgen, T. Jensen, C. Gundlach, L.-B. Tække r, S.V. Hoffman, Z.S. Li, and K. Pedersen. MOS properties of ultra thin oxides on silicon. *Phys. Scripta*, T101:26–29, 2002.

347 S.E. Shaheen, C.J. Brabec, N.S. Sariciftci, F. Padinger, Th. Fromherz, and J.C. Hummelen. 2.5% efficient organic plastic solar cells. *Appl. Phys. Lett.*, 78:841, 2001.

348 A. Goetzberger, B. Voß, and J. Knobloch. *Sonnenenergie: Photovoltaik*. Teubner, Stuttgart, 1997.

349 A. Kay, R. Humphry-Baker, and M. Graetzel. Artificial photosynthesis. 2. investigations on the mechanism of photosensitization of nanocrystalline TiO_2 solar cells by chlorophyll derivatives. *J. Phys. Chem.*, 98:952–959, 1994.

350 A. Hagfeldt and M. Graetzel. Light-induced redox reactions in nanocrystalline systems. *Chem. Rev.*, 95:49–68, 1995.

351 K.M. Kulinowski, P. Jiang, H. Vaswani, and V.L. Colvin. Porous metals from colloidal templates. *Adv. Mater.*, 12:833–838, 2000.

352 P. Jiang, J.F. Bertone, K.S. Hwang, and V.L. Colvin. Single-crystal colloidal multilayers of controlled thickness. *Chem. Mater.*, 11:2132–2140, 1999.

353 A.P. Li, F. Müller, A. Birner, K. Nielsch, and U. Gösele. Hexagonal pore arrays with a 50-420 nm interpore distance formed by self-organization in anodic alumina. *J. Appl. Phys.*, 84:6023–6026, 1998.

354 A. Ashkin. Acceleration and trapping of particles by radiation pressure. *Phys. Rev. Lett.*, 24:156–159, 1970.

355 A. Ashkin, J.M. Dziedzic, J.E. Bjorkholm, and S. Chu. Observation of a single-beam gradient force optical trap for dielectric particles. *Opt. Lett.*, 11:288, 1986.

356 K. Svoboda and S.M. Block. Biological applications of optical forces. *Annu. Rev. Biophys. Biomol. Struct.*, 23:247–285, 1994.

357 A. Ashkin and J.M. Dziedzic AndT. Yamane. Optical trapping and manipulation of single cells using infrared laser beams. *Nature*, 330:769–771, 1987.

358 A. Ashkin and J.M. Dziedzic. Optical trapping and manipulation of viruses and bacteria. *Science*, 235:1517–1520, 1987.

359 S. Chu. Laser manipulation of atoms and particles. *Science*, 253:861–866, 1991.

360 K. Taguchi, K. Atsuta, T. Nakata, and M. Ikeda. Levitation of a microscopic object using plural optical fibers. *Opt. Comm.*, 176:43–47, 2000.

361 M.Righini, A.S.Zelenina, C.Girard, and R.Quidant. Parallel and selective trapping in a patterned plasmonic landscape. *Nature Phys.*, 3:477–480, 2007.

362 F.Ehrenhaft and E.Reeger. *Compt.Rend*, 232:1922, 1951.

363 V.S.Letokhov, V.G.Minogin, and B.D.Pavlik. *Z.Eksp.Teor.Fiz.*, 72:1328, 1977.

364 A. Hemmerich and T.W. Hänsch. Two-dimesional atomic crystal bound by light. *Phys. Rev. Lett.*, 70:410–413, 1993.

365 M. Weidemüller, A. Hemmerich, A. Görlitz, T. Esslinger, and T.W. Hänsch. Bragg diffraction in an atomic lattice bound by light. *Phys. Rev. Lett.*, 75:4583–4586, 1995.

366 M.H. Anderson, J.R. Ensher, M.R. Matthews, C.E. Wieman, and E.A. Cornell. Observation of Bose-Einstein condensation in a dilute atomic vapor. *Science*, 269:198–201, 1995.

367 K.B. Davis, M.-O. Mewes, M.R. Andrews, N.J. van Druten, D.S. Durfee, D.M. Kurn, and W. Ketterle. Bose-Einstein condensation in a gas of sodium atoms. *Phys. Rev. Lett.*, 75:3969–3973, 1995.

368 F. Dalfovo, St. Giorgini, L.P. Pitaevskii, and S. Stringari. Theory of Bose-Einstein condensation in trapped gases. *Rev. Mod. Phys.*, 71:463–512, 1999.

369 S. Martellucci, A.N. Chester, and A. Aspect, editors. *Bose-Einstein Condensates and Atom Lasers*. Kluwer Academic / Plenum Publishers, New York, 2000.

370 E.W.Hagley, L.Deng, W.D.Phillips, K.Burnett, and C.W.Clark. *Opt.Phot.News*, 12(5):22, 2001.

371 M.G. Prentiss. Bound by light. *Science*, 260:1078–1080, 1993.

372 P.K. Ghosh. *Ion Traps*, volume 90 of *International Series of Monographs on Physics*. Oxford University Press, Oxford, 1996.

373 M. Drewsen. private Mitteilung (2003).

374 Y. Yamamoto and R.E. Slusher. Optical processes in microcavities. *Phys. Today*, 46(6):66–73, 1993.

375 P. Berman, editor. *Cavity Quantum Electrodynamics*, Boston, 1994. Academic Press.

376 V. Markel and Th. George. *Optics of Nanostructured Materials*. Wiley Series in Lasers and Applications. Wiley, New York, 2001.

377 M.H. Huang, S. Mao, H. Feick, H. Yan, Y. Wu, H. Kind, E. Weber, R. Russo, and P. Yang. Room-temperature ultraviolet nanowire nanolasers. *Science*, 292:1897–1899, 2001.

378 J.C. Johnson, H. Yan, R.D. Schaller, L.H. Haber, R.J. Saykally, and P. Yang. Single nanowire lasers. *J. Phys. Chem. B*, 105:11387–11390, 2001.

379 F.Quochi, F.Cordella, A.Mura, G.Bongiovanni, F.Balzer, and H.-G.Rubahn. Gain amplification and lasing properties of individual organic nanofibers. *Appl.Phys.Lett.*, 88:041106, 2006.

380 R. Rinaldi, E. Branca, R. Cingolani, S. Masiero, G.P. Spada, and G. Gottarelli. Photodetectors fabricated from a self-assembly of a deoxyguanosine derivative. *Appl. Phys. Lett.*, 78:3541, 2001.

381 A.J. Shields, M.P. O'Sullivan, I. Farrer, D.A. Ritchie, M.L. Leadbeater, N.K. Patel, R.A. Hogg, C.E. Norman, N.J. Curson, and M. Pepper. Single photon detection with a quantum dot transistor. *Jpn. J. Appl. Phys.*, 40:2058–2064, 2001.

382 G. Marowsky, L. Chi, D. Möbius, Y.R. Shen, D. Dorsch, and B. Rieger. Nonlinear optical properties of hemicyanine monolayers and the protonation effect. *Chem. Phys. Lett.*, 147:420–424, 1988.

383 L.R. Dalton, A.W. Harper, and B.H. Robinson. The role of London forces in defining noncentrosymmetric order of high dipole moment-high hyperpolarizability chromophores in electrically poled polymeric thin films. *PNAS*, 94:4842–4847, 1997.

384 A. Yariv. *Optical Electronics*. Holt-Saunders, New York, 1985.

385 H. Fuchs, H. Ohst, and W. Prass. Ultrathin organic films: Molecular architectures for advanced optical, electronic and bio-related systems. *Adv. Mater.*, 3:10–18, 1991.

386 M. Kahl, E. Voges, and W. Hill. Optimization of SERS substrates by electron-beam lithography. *Spectr. Europe*, pages 8–13, October 1998.

387 W. Gotschy, K. Vonmet, A. Leitner, and F.R. Aussenegg. Thin films by regular patterns of metal nanoparticles: Tailoring the optical properties by nanodesign. *Appl. Phys. B*, 63:381–384, 1996.

388 J. Krenn, R. Wolf, A. Leitner, and F. Aussenegg. Near-field optical imaging the surface plasmon fields of lithographically designed nanostructures. *Opt. Communic.*, 137:46–50, 1997.

389 J.R. Krenn, A. Dereux Nd J.C. Weeber, E. Bourillot, Y. Lacroute, J.P. Goudonnet, G. Schider, W. Toschy, A. Leitner, F.R. Aussenegg, and C. Girard. Squeezing the optical near-field zone by plasmon coupling of metallic nanoparticles. *Phys. Rev. Lett.*, 82:2590–2593, 1999.

390 H.S. Nalwa, editor. *Magnetic Nanostructures*. American Scientific Publishers, 2002.

391 M.N. Baibich, J.M. Broto, A. Fert, F.N. Van Dau, F. Petroff, P. Eitenne, G. Creuzet, A. Friederich, and J. Chazelas. Giant magnetoresistance of (001)Fe/(001)Cr magnetic superlattices. *Phys. Rev. Lett.*, 61:2472–2475, 1988.

392 P. Grünberg. Layered magnetic structures: History, highlights, applications. *Phys. Today*, 54(5):31–37, 2001.

393 TeraStor Company, 1999. 930 Wrigley Way, Milpitas, CA 95035, USA.

394 J. Joannopoulos, R. Meade, and J. Winn. *Photonic Crystals*. Princeton Press, Princeton, N.J., 1995.

395 J.D. Joannopoulos, P.R. Villeneuve, and S. Fan. Photonic crystals: Putting a new twist on light. *Nature*, 386:143–149, 1997.

396 J.D. Joannopoulos, P.R. Villeneuve, and S. Fan. Photonic crystals. *Sol. State Comm.*, 102:165–173, 1997.

397 M.M. Sigalas, K.-M. Ho, R. Biswas, and C.M. Soukoulis. Photonic crystals. In V.A. Markel and T.F. George, editors, *Optics of Nanostructured Materials*, New York, 2001. Wiley.

398 E.Yablonovitch. *Scientific American*, page 35, Dec. 2001.

399 S.G. Johnson and J.D. Joannopoulos. Three-dimensionally periodic dielectric layered structure with omnidirectional photonic band gap. *Appl. Phys. Lett.*, 77:3490–3492, 2000.

400 M.Srinivasaro, D.Collings, A.Philips, and S.Patel. *Science*, 292:79, 2001.

401 J.C. Knight, T.A. Birks, P.St.J. Russell, and D.M. Atkin. All-silica single-mode optical fiber with photonic crystal cladding. *Opt. Lett.*, 21:1547, 1996. Errata: Vol. 22, pp.484-485 (1997).

402 J.C. Knight, T.A. Birks, and P.St.J. Russell. "holey" silica fibers. In V. Markel and Th. George, editors, *Optics of Nanostructured Materials*, Wiley Series in Lasers and Applications, pages 39–72, New York, 2001. Wiley.

403 J.K. Ranka, R.S. Windeler, and A.J. Stentz. Visible continuum generation in air-silica microstructure optical fibers with anomalous dispersion at 800 nm. *Opt. Lett.*, 25:25–27, 2000.

404 W.J. Wadsworth, J.C. Knight, A. Ortigosa-Blanch, J. Arriaga, E. Silvestre, and P.St.J. Russell. Soliton effects in photonic crystal fibres at 850 nm. *Electron. Lett.*, 36(1):53–55, 2000.

405 J.-C. Diels and W. Rudolph. *Ultrafast Laser Pulse Phenomena*. Academic Press, San Diego, 1996.

406 J.Boness and H.-G.Rubahn. *Physik in unserer Zeit*, 31:121, 2000.

407 C. Rulliere, editor. *Femtosecond Laser Pulses*. Springer, Berlin, 2004.

408 J.-H. Kleinwiele. Diplomarbeit, University of Göttingen, 1997.

409 J.-C. Diels, J. Fontaine, I. McMichael, and F. Simoni. Control and measurement of ultrashort pulse shapes (in amplitude and phase) with femtosecond accuracy. *Appl. Opt.*, 24:1270–1282, 1985.

410 M.V. Exter and A. Lagendijk. Ultrashort surface-plasmon and phonon dynamics. *Phys. Rev. Lett.*, 60:49–52, 1988.

411 N. Kroo, W. Krieger, Z. Lenkefi, Z. Szentirmay, J. Thost, and H. Walther. A new optical method for investigation of thin metal films. *Surf. Sci.*, 331-333:1305–1309, 1995.

412 N. Kroo and Z. Szentirmay. Decay time measurement of surface plasmons on silver gratings. *Hung. Acad. Sci. KFKI*, 1988-18/E:1:1–13, 1988.

413 W. Wang, M. Feldstein, and N. Scherer. Observation of coherent multiple scattering of surface plasmon polaritons on Ag and Au surfaces. *Chem. Phys. Lett.*, 262:573–582, 1996.

414 J. Hohlfeld, U. Conrad, and E. Matthias. Does femtosecond time-resolved second-harmonic generation probe electron temperatures at surfaces? *Appl. Phys. B*, 63:541–544, 1996.

415 T. Luce, W. Hübner, and K. Bennemann. Theory for the nonlinear optical response at noble-metal surfaces with nonequilibrium electrons. *Z. Phys. B*, 102:223–232, 1997.

416 C.-K. Sun, F. Vallee, L.H. Acioli, E.P. Ippen, and J.G. Fujimoto. Femtosecond-tunable measurement of electron thermalization in gold. *Phys. Rev. B*, 50:15337–15348, 1994.

417 S. Anisimov, B. Kapeliovich, and T. Perel'man. Emission of electrons from the surface of metals induced by ultrashort laser pulses. *Sov. Phys. JETP*, 66:776–781, 1974.

418 R. Groeneveld, R. Sprik, and A. Lagendijk. Femtosecond spectroscopy of electron-electron and electron-phonon energy relaxation in Ag and Au. *Phys. Rev. B*, 51:11433–11445, 1995.

419 S. Brorson, J. Fujimoto, and E. Ippen. Femtosecond electronic heat-transport dynamics in thin gold films. *Phys. Rev. Lett.*, 59:1962–1965, 1987.

420 J. Hohlfeld, J. Müller, S.-S. Wellershof, and E. Matthias. Time-resolved thermoreflectivity of thin gold films and its dependence on film thickness. *Appl. Phys. B*, 64:387–390, 1997.

421 R.W. Schoenlein, J.G. Fujimoto, G.L. Eesley, and T.W. Capehart. Femtosecond studies of image-potential dynamics in metals. *Phys. Rev. Lett.*, 61:2596–2599, 1988.

422 T. Fauster and W. Steinmann. Two-photon photoemission spectroscopy of image states. In P. Halevi, editor, *Photonic Probes of Surfaces*, pages 347–411, Amsterdam, 1995. Elsevier.

423 R. L. Lingle, Jr., D. F. Padowitz, R. E. Jordan, J. D. McNeill, and C. B. Harris. Two-dimensional localization of electrons at interfaces. *Phys. Rev. Lett.*, 72:2243–2246, 1994.

424 N.-H. Ge, C. Wong, R. Lingle, J. McNeill, K. Gaffney, and C. Harris. Femtosecond dynamics of electron localization at interfaces. *Science*, 279:202–205, 1998.

425 M. Wolf. Femtosecond dynamics of electronic excitations at metal surfaces. *Surf. Sci.*, 377 - 379:343–349, 1997.

426 W. Fann, R. Storz, H. Tom, and J. Bokor. Electron thermalization in gold. *Phys. Rev. B*, 46:13592–13595, 1992.

427 R.L. Lingle, Jr., N.-H. Ge, R.E. Jordan, J.D. McNeill, and C.B. Harris. Femtosecond studies of electron tunneling at metal-dielectric interfaces. *Chem. Phys.*, 205:191–203, 1996.

428 C.A. Schmuttenmaer, M. Aeschlimann, H.E. Elsayed-Ali, R.J.D. Miller, D.A. Mantell, J. Cao, and Y. Gao. Time-resolved two-photon photoemission from Cu(100): Energy dependence of electron relaxation. *Phys. Rev. B*, 50:8957–8960, 1994.

429 T. Hertel, E. Knoesel, M. Wolf, and G. Ertl. Ultrafast electron dynamics at Cu(111): Response of an electron gas to optical excitation. *Phys. Rev. Lett.*, 76:535–538, 1996.

430 H. Petek and S. Ogawa. Femtosecond time-resolved two-photon photoemission studies of electron dynamics in metals. *Progr. Surf. Sci.*, 56:239–310, 1997.

431 M. Bauer, S. Pawlik, and M. Aeschlimann. Resonance lifetime and energy of an excited Cs state on Cu(111). *Phys. Rev. B*, 55:10040–10043, 1997.

432 E. Knoesel, A. Hotzel, and M. Wolf. Temperature dependence of surface state lifetimes, dephasing rates and binding energies on Cu(111) studied with time-resolved photoemission. *J. Electron. Spectrosc. Rel. Phen.*, 88 - 91:577–584, 1998.

433 S. Ogawa, H. Nagano, H. Petek, and A.P. Heberle. Optical dephasing in Cu(111) measured by interferometric two-photon time-resolved photoemission. *Phys. Rev. Lett.*, 78:1339–1342, 1997.

434 L. Bartels, G. Meyer, K.-H. Rieder, D. Velic, E. Knoesel, A. Hotzel, M. Wolf, and G. Ertl. Dynamics of electron-induced manipulation of individual CO molecules on Cu(111). *Phys. Rev. Lett.*, 80:2004–2007, 1998.

435 M. Shapiro and P. Brunner. Quantum control of chemical reactions. *J. Chem. Soc. Faraday Trans.*, 93:1263–1277, 1997.

436 H. van Driel, J. Sipe, A. Hache, and R. Atanasov. Coherence control of photocurrents in semiconductors. *Phys. Status Solidi B*, 204:3–8, 1997.

437 H. Petek, A.P. Heberle, W. Nessler, H. Nagano, S. Kubota, S. Matsunami, N. Moriya, and S. Ogawa. Optical phase control of coherent electron dynamics in metals. *Phys. Rev. Lett.*, 79:4649–4652, 1997.

438 X.-P. Jiang, M. Shapiro, and P. Brumer. Electronic absorption spectroscopy of diatomics on a dynamic surface: IBr on MgO(001). *J. Chem. Phys.*, 105:3479–3485, 1996.

439 C. Voisin, N. Del Fatti, D. Christofilos, and F. Vallee. Ultrafast electron dynamics and optical nonlinearities in metal nanoparticles. *J. Phys. Chem. B*, 105:2264–2280, 2001.

440 T. Schröder, R. Schinke, R. Krohne, and U. Buck. Vibrational dynamics of large clusters from helium atom scattering: Calculations for ar_{55}. *J. Chem. Phys.*, 106:9067–9077, 1997.

441 H. Shinojima, J. Yumoto, and N. Uesugi. Size dependence of optical nonlinearity of CdSSe microcrystallites doped in glass. *Appl. Phys. Lett.*, 60:298–300, 1992.

442 H. Inouye, K. Tanaka, I. Tanahashi, and K. Hirao. Ultrafast dynamics of nonequilibrium electrons in a gold nanoparticle system. *Phys. Rev. B*, 57:11334–11340, 1998.

443 J.H. Hodak, I. Martini, and G.V. Hartland. Spectroscopy and dynamics of nanometer-sized noble metal particles. *J. Phys. Chem. B*, 102:6958–6967, 1998.

444 G. Mie. Beiträge zur Optik trüber Medien speziell kolloidaler Metallösungen. *Ann. Phys. (Leipzig)*, 25:377–445, 1908.

445 U. Kreibig and M. Vollmer. *Optical Properties of Metal Clusters*, volume 25 of *Springer Series in Materials Science*. Springer-Verlag, Berlin, 1995.

446 N.D. Lang and W. Kohn. Theory of metal surfaces: Charge density and surface energy. *Phys. Rev. B*, 1:4555–4568, 1970.

447 C.F Bohren and D.R. Huffman. *Absorption and Scattering of Light by Small Particles*. John Wiley & Sons, New York, 1983.

448 P.W. Barber and S.C. Hill. *Light Scattering by Particles: Computational Methods*, volume 2 of *Advanced Series in Applied Physics*. World Scientific, Singapore, 1990.

449 M. Rasigni, G. Rasigni, J.P. Gasparini, and R. Fraisse. Structure and optical conductivity of thin lithium deposits prepared at 6 K. *J. Appl. Phys.*, 47:1757–1761, 1976.

450 H. Schmeisser. Growth and mobility effects of gold clusters on rocksalt (100) surfaces studied with the method of quantitative image analysis. part I: Cluster size distributions. *Thin Solid Films*, 22:83–97, 1974.

451 P. Royer, J.L. Bijeon, J.P. Goudonnet, T. Inagaki, and E.T. Arakawa. Optical absorbance of silver oblate particles. *Surface Science*, 217:384–402, 1989.

452 R.R. Singer, A. Leitner, and F.R. Aussenegg. Structure analysis and models for optical constants of discontinuous metallic silver films. *J. Opt. Soc. Am. B*, 12:220–228, 1995.

453 U. Kreibig, M. Gartz, and A. Hilger. Mie resonances: Sensors for physical and chemical cluster interface properties. *Ber. Bunsenges. Phys. Chem.*, 101:1–12, 1997.

454 U. Kreibig. Optics of nanosized metals. In R.E. Hummel and P. Wimann, editors, *Handbook of Optical Properties, Vol. II, Optics of Small Particles, Interfaces and Surfaces*, page 145, Boca Raton, 1997. CRC Press.

455 J.-H. Klein-Wiele, P. Simon, and H.-G. Rubahn. Size-dependent plasmon lifetimes and electron-phonon coupling time constants for surface bound Na clusters. *Phys. Rev. Lett.*, 80:45–48, 1998.

456 B. Lamprecht, A. Leitner, and F.R. Aussenegg. SHG studies of plasmon dephasing in nanoparticles. *Appl. Phys. B*, 68:419–423, 1999.

457 W.P. Halperin. Quantum size effects in metal particles. *Rev. Mod. Phys.*, 58:533–606, 1986.

458 C. Yannouleas, E. Vigezzi, and R.A. Broglia. Evolution of the optical properties of alkali-metal microclusters towards the bulk: The matrix random-phase approximation description. *Phys. Rev. B*, 47:9849–9861, 1993.

459 W.C. Huang and J.T. Lue. Quantum size effect on the optical properties of small metallic particles. *Phys. Rev. B*, 49:17279–17285, 1994.

460 C.A. Ullrich and G. Vignale. Linewidths of collective excitations of the inhomogeneous electron gas: Application to two-dimensional quantum strips. *Phys. Rev. B*, 58:7141–7150, 1998.

461 G.F. Bertsch, N. Van Giai, and N. Vinh Mau. Cluster ionization via two-plasmon excitation. *Phys. Rev. A*, 61:033202, 2000.

462 R. Schlipper, R. Kusche, B. von Issendorff, and H. Haberland. Multiple excitation and lifetime of the sodium cluster plasmon resonance. *Phys. Rev. Lett.*, 80:1194–1197, 1998.

463 J. Lehmann, M. Merschdorf, W. Pfeifer, A. Thon, S. Voll, and G. Gerber. Surface plasmon dynamics in silver nanoparticles studied by femtosecond time-resolved photoemission. *Phys. Rev. Lett.*, 85:2921–2924, 2000.

464 J.H. Parks and S.A. McDonald. Evolution of the collective-mode resonance in small adsorbed sodium clusters. *Phys. Rev. Lett.*, 62:2301–2304, 1989.

465 T. Klar, M. Perner, S. Grosse, G. von Plessen, W. Spirkl, and J. Feldmann. Surface-plasmon resonances in single metallic nanoparticles. *Phys. Rev. Lett.*, 80:4249–4252, 1998.

466 N. Nilius, N. Ernst, and H.-J. Freund. Photon emission spectroscopy of individual oxide-supported silver clusters in a scanning tunneling microscope. *Phys. Rev. Lett.*, 84:3994–3997, 2000.

467 N. Nilius, N. Ernst, and H. J. Freund. Photon emission from individual supported gold clusters: Thin film versus bulk oxide. *Surface Science*, 478:L327–L332, 2001.

468 C. Sönnichsen, S. Geier, N.E. Hecker, G. von Plessen, J. Feldmann, H. Ditlbacher, B. Lamprecht, J.R. Krenn, F.R. Aussenegg, V.Z.-H. Chan, J.P. Spatz, and M. Möller. Spectroscopy of single metallic nanoparticles using total internal reflection microscopy. *Appl. Phys. Lett.*, 77:2949–2951, 2000.

469 D. Steinmüller-Nethl, R. A. Höpfel, E. Gornik, A. Leitner, and F.R. Aussenegg. Femtosecond relaxation of localized plasma excitations in Ag islands. *Phys. Rev. Lett.*, 68:389–392, 1992.

470 B. Lamprecht, A. Leitner, and F.R. Aussenegg. Femtosecond decay-time measurement of electron plasma oscillation in nanolithographically designed silver particles. *Appl. Phys. B*, 64:269–272, 1997.

471 B. Lamprecht, J.R. Krenn, A. Leitner, and F.R. Aussenegg. Particle-plasmon decay-time determination by measuring the optical near-field's autocorrelation: Influence of inhomogeneous line broadening. *Appl. Phys. B*, 69:223–227, 1999.

472 H.J. Simon, D.E. Mitchell, and J.G. Watson. Second harmonic generation with surface plasmons in alkali metals. *Opt. Comm.*, 13:294–298, 1975.

473 N. Papadogiannis, S. Moustaizis, P. Loukakos, and C. Kalpouzos. Temporal characterization of ultra short laser pulses based on multiple harmonic generation on a gold surface. *Appl. Phys. B*, 65:339–345, 1997.

474 L. Cruz, L.F. Fonseca, and M. Gómez. T-matrix approach for the calculation of local fields in the neighborhood of small clusters in the electrodynamic regime. *Phys. Rev. B*, 40:7491–7500, 1989.

475 D. Pines and P. Nozieres. *The Theory of Quantum Liquids*. Benjamin, New York, 1966.

476 H.-G. Rubahn. Time constants for the decay of elementary optical excitations in surface bound Na clusters. *Appl. Surf. Sci.*, 109/110:575–578, 1997.

477 J.-H. Klein-Wiele, P. Simon, and H.-G. Rubahn. Picosecond response of sodium clusters on dielectric substrates. *Opt. Comm.*, 161:42–46, 1999.

478 H.E. Elsayed-Ali and T. Juhasz. Femtosecond time-resolved thermomodulation of thin gold films with different crystal structures. *Phys. Rev. B*, 47:13599–13610, 1993.

479 C. Suarez, W.E. Bron, and T. Juhasz. Dynamics and transport of electronic carriers in thin gold films. *Phys. Rev. Lett.*, 75:4536–4539, 1995.

480 A. Zewail. *Femtochemistry: Ultrafast Dynamics of the Chemical Bond*. World Scientific, Singapore, 1994.

481 B. Stipe, M. Rezaei, and W. Ho. Inducing and viewing the rotational motion of a single molecule. *Science*, 279:1907–1909, 1998.

482 M. Feldstein, P. Vöhringer, W. Wang, and N. Scherer. Femtosecond optical spectroscopy and scanning probe microscopy. *J. Phys. Chem.*, 100:4739–4748, 1996.

483 M. Feldstein and N. Scherer. Femtosecond correlated optical reactivity and scanning tunneling microscopy studies on metal surfaces. *Proc. SPIE*, 3272:58–66, 1998.

484 D. von der Linde, K. Sokolowski-Tinten, and J. Bialkowski. Laser-solid interaction in the femtosecond time regime. *Appl. Surf. Sci.*, 109 - 110:1–10, 1997.

485 G. Mourou and S. Williamson. Picosecond electron diffraction. *Appl. Phys. Lett.*, 41:44–45, 1982.

486 M. Aeschlimann, E. Hull, J. Cao, C. Schmuttenmaer, L. Jahn, Y. Gao, H. Elsayed-Ali, D. Mantell, and M. Scheinfein. A picosecond electron gun for surface analysis. *Rev. Sci. Instrum.*, 66:1000–1009, 1995.

487 H.E. Elsayed-Ali and J.W. Herman. Ultrahigh vacuum picosecond laser-driven electron diffraction system. *Rev. Sci. Instrum.*, 61:1636–1647, 1990.

488 J.W.M. Frenken and J.F. van der Veen. Observation of surface melting. *Phys. Rev. Lett.*, 54:134–137, 1985.

489 J.W. Herman and H.E. Elsayed-Ali. Time-resolved study of surface disordering of Pb(110). *Phys. Rev. Lett.*, 68:2952–2955, 1992.

490 J.W. Herman and H.E. Elsayed-Ali. Superheating of Pb(111). *Phys. Rev. Lett.*, 69:1228–1231, 1992.

491 J.W. Herman, H.E. Elsayed-Ali, and E.A. Murphy. Time-resolved structural study of Pb(100). *Phys. Rev. Lett.*, 71:400–403, 1993.

492 L. Larsson, P. Heimann, A. Lindenberg, P. Schuck, P. Bucksbaum, R. Lee, H. Padmore, J. Wark, and R. Falcone. Ultrafast structural changes measured by time-resolved X-ray diffraction. *Appl. Phys. A*, 66:587–591, 1998.

493 X. Zeng, B. Lin, I. El-Kholy, and H.E. Elsayed-Ali. Time-resolved reflection high-energy electron diffraction study of the Ge(111)-c(2×8)-(1×1) phase transition. *Phys. Rev. B*, 59:14907–14910, 1999.

494 X. Zeng, B. Lin, I. El-Kholy, and H.E. Elsayed-Ali. Time-resolved structural study of the Ge(111) high-temperature phase transition. *Surf. Sci.*, 439:95–102, 1999.

495 X. Zeng and H.E. Elsayed-Ali. Surface thermal expansion of Ge(111). *Surf. Sci. Lett.*, 442:L977–L982, 1999.

496 J. Williamson, J. Cao, H. Ihee, H. Frey, and A. Zewail. Clocking transient chemical changes by ultrafast electron diffraction. *Nature*, 386:159–161, 1997.

497 H. Ihee, V.A. Lobastov, U.M. Gomez, B.M. Goodson, R. Srinivasan, C.-Y. Ruan, and A.H. Zewail. Direct imaging of transient molecular structures with ultrafast diffraction. *Science*, 291:458–462, 2001.

498 A. Aviram and M.A. Ratner. Molecular rectifiers. *Chem. Phys. Lett.*, 29:277–283, 1974.

499 E. Scheer, N. Agrait, J.C. Cuevas, A.L. Yeyati, B. Ludoph, A. Martin-Rodero, G.R. Bollinger, J.M. van Ruitenbeek, and C. Urbina. The signature of chemical valence in the electrical conduction through a single-atom contact. *Nature*, 394:154–157, 1998.

500 S. Hong, R. Reifenberger, W. Tian, S. Datta, J. Henderson, and C.P. Kubiak. Molecular conductance spectroscopy of conjugated, phenyl-based molecules on Au(111): The effect of end groups on molecular conduction. *Superlattices and Microstructures*, 28:289–303, 2000.

501 H.X. He, C.Z. Li, and N.J. Tao. Conductance of polymer nanowires fabricated by a combined electrodeposition and mechanical break junction method. *Appl. Phys. Lett.*, 78:811–813, 2001.

502 E. Braun, Y. Eichen, U. Sivan, and G. Ben-Yoseph. DNA-templated assembly and electrode attachment of a conducting silver wire. *Nature*, 391:775–778, 1998.

503 M.A. Rampi, O.J.A. Schueller, and G.M. Whiteside. Alkanethiol self-assembled monolayers as the dielectric of capacitors with nanoscale thickness. *Appl. Phys. Lett.*, 72:1781–1783, 1998.

504 C. Zhou, M.R. Deshpande, M.A. Reed, L. Jones II, and J.M. Tour. Nanoscale metal/self-assembled monolayer/metal heterostructures. *Appl. Phys. Lett.*, 71:611–613, 1997.

505 V.G. Bordo F. Balzer and H.-G. Rubahn. Surface-induced changes of the optical response of particles in nanoscaled systems: A combined experimental and theoretical study. *SPIE Proceedings*, 3272:42–50, 1998.

506 F. Balzer, S.D. Jett, and H.-G. Rubahn. Non-linear optically active metal clusters in nanoscaled systems including self-assembled organic films. *Thin Solid Films*, 372:78–84, 2000.

507 F. Balzer and H.-G. Rubahn. Second-harmonic generation and shielding effects of alkali clusters on ultrathin organic films. *Nanotechnology*, 12:105–109, 2001.

508 Th. Müller, P.H. Vaccaro, F. Balzer, and H.-G. Rubahn. Size dependent optical second harmonic generation from surface bound Na clusters: Comparison between experiment and theory. *Opt. Comm.*, 135:103–108, 1997.

509 C.B.Murray, C.R.Kagan, and M.G.Bawendi. *Science*, 270:1335, 1996.

510 D.L. Klein, R. Roth, A.K.L. Lim, A.P. Alivisatos, and P.L. McEuen. A single-electron transistor made from a cadmium selenide nanocrystal. *Nature*, 389:699–701, 1997.

511 C. Schoessler and H.W.P. Koops. Nanostructured integrated electron source. *J. Vac. Sci. Technol. B*, 16:862–865, 1998.

512 R.P. Feynman. Simulating physics with computers. *Int. J. Theor. Phys.*, 21:467–488, 1982.

513 D. Deutsch. Quantum theory, the Church-Turing principle and the universal quantum computer. *Proc. R. Soc. A*, 400:97–117, 1985.

514 D.J. Wineland, R.E. Drullinger, and F.L. Walls. Radiation-pressure cooling of bound resonant absorbers. *Phys. Rev. Lett.*, 40:1639–1642, 1978.

515 W. Neuhauser, M. Hohenstatt, P. Toschek, and H. Dehmelt. Optical-sideband cooling of visible atom cloud confined in parabolic well. *Phys. Rev. Lett.*, 41:233–236, 1978.

516 J. Dalibard and C. Cohen-Tannoudji. Laser cooling below the Doppler limit by polarization gradients: Simple theoretical models. *J. Opt. Soc. Am. B*, 6:2023, 1989.

517 G. Morigi, J. Eschner, J.I. Cirac, and P. Zoller. Laser cooling of two trapped ions: Sideband cooling beyond the Lamb-Dicke limit. *Phys. Rev. A*, 59:3797–3808, 1999.

518 J.I. Cirac and P. Zoller. Quantum computations with cold trapped ions. *Phys. Rev. Lett.*, 74:4091–4094, 1995.

519 C. Monroe, D.M. Meekhof, B.E. King, S.R. Jefferts, W.M. Itano, D.J. Wineland, and P. Gould. Resolved-sideband Raman cooling of a bound atom to the 3D zero-point energy. *Phys. Rev. Lett.*, 75:4011–4014, 1995.

520 L.M. K. Vandersypen, M. Steffen, G. Breyta, C.S. Yannoni, M.H. Sherwood, and I.L. Chuang. Experimental realization of Shor's quantum factoring algorithm using nuclear magnetic resonance. *Nature*, 414:883–887, 2001.

521 C.A. Sackett, D. Kielpinski, B.E. King, C. Langer, V. Meyer, C.J. Myatt, M. Rowe, Q.A. Turchette, W.M. Itano, D.J. Wineland, and C. Monroe. Experimental entanglement of four particles. *Nature*, 404:256–259, 2000.

522 K.D. Jandt. Atomic force microscopy of biomaterials surfaces and interfaces. *Surf. Sci.*, 491:303–332, 2001.

523 H. Fang, L.C. Giancarlo, and G.W. Flynn. Direct determination of the chirality of organic molecules by scanning tunneling microscopy. *J. Phys. Chem. B*, 102:7311–7315, 1998.

524 A. Kühnle, T.R. Linderoth, B. Hammer, and F. Besenbacher. Chiral recognition in dimerization of adsorbed cysteine observed by scanning tunnelling microscopy. *Nature*, 415:891–893, 2002.

525 T. Tojima, D. Hatakeyama, Y. Yamane, K. Kawabata, T. Ushiki, S. Ogura, K. Abe, and E. Ito. Comparative atomic force and scanning electron microscopy for fine structural images of nerve cells. *Jpn. J. Appl. Phys.*, 37:3855–3859, 1998.

526 H. Clausen-Schaumann, M. Seitz, R. Krautbauer, and H.E. Gaub. Force spectroscopy with single biomolecules. *Curr. Op. Chem. Biolog.*, 4:524–530, 2000.

527 A. Janshoff, M. Neitzert, Y. Oberdörfer, and H. Fuchs. Kraftspektroskopie an molekularen Systemen - Einzelmolekülspektroskopie an Polymeren und Biomolekülen. *Angew. Chem.*, 112:3346–3374, 2000.

528 T. Strick, J.-F. Allemand, V. Croquette, and D. Bensimon. The manipulation of single biomolecules. *Phys. Today*, 54(10):46–51, 2001.

529 V. Balzani, A. Credi, F.M. Raymo, and J.F. Stoddart. Künstliche molekulare Maschinen. *Angew. Chem.*, 112:3484–3530, 2000.

530 R. Krautbauer, H. Clausen-Schaumann, and H.E. Gaub. Cisplatin verändert die Mechanik von DNA-Molekülen. *Angew. Chem.*, 112:4056–4059, 2000.

531 M. Grandbois, M. Beyer, M. Rief, H. Clausen-Schaumann, and H.E. Gaub. How strong is a covalent bond? *Science*, 283:1727–1730, 1999.

532 H. Clausen-Schaumann, M. Rief, C. Tolksdorf, and H.E. Gaub. Mechanical stability of single DNA molecules. *Biophys. J.*, 78:1997–2007, 2000.

533 W. Nachtigall. *Bionik*. Springer, Berlin, 2002.

534 M.Srinivasaro. *Chem.Rev.*, 99:1935, 1999.

535 V.C. Sundar, A.D. Yablon, J.L. Grazul, M. Ilan, and J. Aizenberg. Fibre-optical features of a glass sponge. *Nature*, 424:899–900, 2003.

536 J. Aizenberg, A. Tkachenko, S. Weiner, L. Addadi, and G. Hendler. Calcitic microlenses as part of the photoreceptor system in brittlestars. *Nature*, 412:819–822, 2001.

537 M. Gross. *Travels to the Nanoworld: Miniature Machinery in Nature and Technology*. Perseus Books, Cambridge, 2001.

538 J.J. Storhoff and C.A. Mirkin. Programmed materials synthesis with DNA. *Chem. Rev.*, 99:1849–1862, 1999.

539 M. Mertig, R. Kirsch, W. Pompe, and H. Engelhardt. Fabrication of highly oriented nanocluster arrays by biomolecular templating. *EPJ D*, 9:45–48, 1999.

540 F.D. Lewis, T. Wu, Y. Zhang, R.L. Letsinger, S.R. Greenfield, and M.R. Wasielewski. Distance-dependent electron transfer in DNA hairpins. *Science*, 277:673–676, 1997.

541 J.Baggot. *Perfect Symmetry: The Accidental Discovery of Buckminsterfullerene*. Oxford University Press, Oxford, 1994.

542 H. Aldersey-Williams. *The Most Beautiful Molecule: The Discovery of the Buckyball*. John Wiley, New York, 1997.

543 S. Dresselhaus, G. Dresselhaus, and P.C. Eklund. *Science of Fullerenes and Carbon Nanotubes: Their Properties and Applications*. Academic Press, San Diego, 1996.

544 S. Iijima. Helical microtubules of graphitic carbon. *Nature*, 354:56–58, 1991.

545 E.W. Wong, P.E. Sheehan, and C.M. Lieber. Nanobeam mechanics: Elasticity, strength, and toughness of nanorods and nanotubes. *Science*, 277:1971–1975, 1997.

546 P.J.F. Harris. *Carbon Nanotubes and Related Structures. New Materials for the Twenty-First Century*. Cambridge University Press, Cambridge, 2002.

547 M.J. Rosker, H.O. Marcy, T.Y. Chang, J.T. Khoury, K. Hansen, and R.L. Whetten. Time-resolved degenerate four-wave mixing in thin films of C_{60} and C_{70} using femtosecond optical pulses. *Chem. Phys. Lett.*, 196:427–432, 1992.

548 A. Thess, R. Lee, P. Nikolaev, H. Dai, P. Petit, J. Robert, C. Xu, Y.H. Lee, S.G. Kim, A.G. Rinzler, D.T. Colbert, G.E. Scuseria, D. Tománek, J.E. Fischer, and R.E. Smalley. Crystalline ropes of metallic carbon nanotubes. *Science*, 273:483–487, 1996.

549 P.M. Ajayan. Nanotubes from carbon. *Chem. Rev.*, 99:1787–1800, 1999.

550 S.J. Tans, M.H. Devoret, H. Dai, A. Thess, R.E. Smalley, L.J. Geerligs, and C. Dekker. Individual single-wall carbon nanotubes as quantum wires. *Nature*, 386:474–477, 1997.

551 J.N. Coleman, S. Curran, A.B. Dalton, A.P. Davey, B. Mc Carthy, W. Blau, and R.C. Barklie. Physical doping of a conjugated polymer with carbon nanotubes. *Synth. Met.*, 102:1174–1175, 1999.

552 S.J. Tans, A.R.M. Verschueren, and C. Dekker. Room-temperature transistor based on a single carbon nanotube. *Nature*, 393:49–52, 1998.

553 O. Gröning, O.M. Küttel, Ch. Emmenegger, P. Gröning, and L. Schlapbach. Field emission properties of carbon nanotubes. *J. Vac. Sci. Technol. B*, 18:665–678, 2000.

554 H. Kind, J.-M. Bonard, C. Emmenegger, L.-O. Nilsson, K. Hernadi, E. Maillard-Schaller, L. Schlapbach, L. Forró, and K. Kern. Patterned films of nanotubes using microcontact printing of catalysts. *Adv. Mater.*, 11:1285–1289, 1999.

555 L. Nilsson, O. Gröning, C. Emmenegger, O. Kuettel, E. Schaller, L. Schlapbach, H. Kind, J.-M. Bonard, and K. Kern. Scanning field emission from patterned carbon nanotube films. *Appl. Phys. Lett.*, 76:2071–2073, 2000.

556 B. Bhushan, editor. *Micro/Nanotribology and its Applications*, volume 330 of *NATO Science Series E:*. Kluwer, Dordrecht, 1997.

557 T. Zijlstra, J.A. Heimberg, E. van der Drift, D. Glastra van Loon, M. Dienwiebel, L.E.M. de Groot, and J.W.M. Frenken. Fabrication of a novel scanning probe device for quantitative nanotribology. *Sens. Act. A*, 84:18–24, 2000.

558 P. Rai-Choudhury, editor. *MEMS and MOEMS Technology and Applications*. SPIE Press, Bellingham, 2000.

559 D. Bishop, C.R. Giles, and P. Gammel. The little machines that are making it big. *Phys. Today*, 54(10):38–44, 2001.

560 D. Dragoman and M. Dragoman. Micro/nano-optoeletromechanical systems. *Progr. Quant. Electr.*, 25:229–290, 2001.

561 S.R. Quake and A. Scherer. From micro- to nanofabrication with soft materials. *Science*, 290:1536–1540, 2000.

562 P. Kim and C.M. Lieber. Nanotube nanotweezers. *Science*, 286:2148–2150, 1999.

563 S. Akita, Y. Nakayama, S. Mizooka, Y. Takano, Okawa T, Y. Miyatake, S. Yamanaka, M. Tsuji, and T. Nosaka. Nanotweezers consisting of carbon nanotubes operating in an atomic force microscope. *Appl. Phys. Lett.*, 79:1691–1693, 2001.

564 J.N. Israelachvili. *Intermolecular and Surface Forces*. Academic Press, London, 2nd edition, 1992.

565 Q. Niu, M.C. Chang, and C.K. Shih. Double-tip scanning tunneling microscope for surface analysis. *Phys. Rev. B*, 51:5502–5505, 1995.

566 P. Bøggild, T.M. Hansen, C. Tanasa, and F. Grey. Fabrication and actuation of customized nanotweezers with a 25 nm gap. *Nanotechn.*, 12:331–335, 2001.

567 F. Besenbacher and J.K. Nørskov. How to power a nanomotor. *Science*, 290:1520, 2000.

568 A.K. Schmid, N.C. Bartelt, and R.Q. Hwang. Alloying at surfaces by the migration of reactive two-dimensional islands. *Science*, 290:1561–1564, 2000.

569 G. Schill. *Catenanes, Rotaxanes, Knots*. Organic Chemistry Monographs. Academic Press, New York, 1971.

570 J.-P. Sauvage and C. Dietrich-Buchecker, editors. *Molecular Catenanes, Rotaxanes and Knots*. Wiley-VCH, Weinheim, 1999.

571 T. Elston, H. Wang, and G. Oster. Energy transduction in ATP synthase. *Nature*, 391:510–513, 1998.

572 D.S. Goodsell. *Our Moleculare Nature. The Body's Motors, Machines and Messages*. Springer, Berlin, 2002.

573 J.R. Dennis, J. Howard, and V. Vogel. Molecular shuttles: Directed motion of microtubules along nanoscale kinesin tracks. *Nanotechn.*, 10:232–236, 1999.

574 N. Koumura, R.W.J. Zijlstra, R.A. van Delden, N. Harada, and B.L. Feringa. Light-driven monodirectional molecular rotor. *Nature*, 401:152–155, 1999.

575 T.R. Kelly, H. De Silva, and R.A. Silva. Unidirectional rotary motion in a molecular system. *Nature*, 401:150–152, 1999.

Index

Basics of Nanotechnology: 3rd Edition. Horst-Günter Rubahn

ISBN: 978-3-527-40800-9